Microbiomes and Emerging Applications

Multidisciplinary Applications and Advances in Biotechnology: Contributions from the Biotechnology Society of Nigeria Working Groups

Series Editors:

Benjamin Ewa Ubi

Department of Biotechnology / Biotechnology Research & Development Centre, Ebonyi State University Abakaliki, Nigeria

Abdulrazak B. Ibrahim

Department of Biochemistry, Ahmadu Bello University, Zaria, Nigeria / Forum for Agricultural Research in Africa (FARA), Accra, Ghana

Nwadiuto (Diuto) Esiobu

Microbial Biotechnology Laboratory, Florida Atlantic University, Boco Raton, FL., USA

Applied Biotech International Nigeria Ltd, Wuye, Abuja, Nigeria

James Chukwuma Ogbonna

Department of Microbiology, University of Nigeria, Nsukka, Nigeria

Charles Oluwaseun Adetunji

Edo State University Uzairue, Nigeria

Olawole Obembe

UNESCO Chair on Plant Biotechnology, Plant Science Research Cluster, Department of Biological Science, Covenant University Ota, Nigeria

Ifeoma M. Ezeonu

Department of Microbiology

University of Nigeria Nsukka, Nigeria

Fermentation and Algal Biotechnologies for the Food, Beverage and Other Bioproduct Industries

James Chukwuma Ogbonna, Sylvia Uzochukwu, Emeka Godfrey Nwoba, Charles Oluwaseun Adetunji, Nwadiuto (Diuto) Esiobu, Abdulrazak B. Ibrahim, Benjamin Ewa Ubi

Agricultural Biotechnology, Biodiversity and Bioresources Conservation and Utilization

Olawole Obembe, Emmanuel Olufemi Ekundayo, Arinze Stanley Okoli, Abubakar Gidado, Charles Oluwaseun Adetunji, Abdulrazak B. Ibrahim, Benjamin Ewa Ubi

Medical Biotechnology, Biopharmaceutics, Forensic Science and Bioinformatics
Hajia Mairo Inuwa, Ifeoma Maureen Ezeonu, Charles Oluwaseun Adetunji, Abubakar Gidado, Emmanuel Olufemi Ekundayo, Abdulrazak B. Ibrahim, Benjamin Ewa Ubi

Biosafety and Bioethics in Biotechnology: Policy, Advocacy, and Capacity Building
Sylvia Uzochukwu, Okoli Arinze Stanley Okoli, Nwadiuto (Diuto) Esiobu, Emeka Godfrey Nwoba, Christpeace Nwagbo Ezebuiro, Charles Oluwaseun Adetunji, Abdulrazak Ibrahim, and Benjamin Ewa Ubi

Microbiomes and Emerging Applications
Nwadiuto (Diuto) Esiobu, James Chukwuma Ogbonna, Charles Oluwaseun Adetunji, Olawole Obembe, Ifeoma Maureen Ezeonu, Abdulrazak Ibrahim, and Benjamin Ewa Ubi

Bioenergy and Environmental Biotechnology for Sustainable Development
Akinola Rasheed Popoola, Emeka Godfrey Nwoba, James Chukwuma Ogbonna, Charles Oluwaseun Adetunji, Nwadiuto (Diuto) Esiobu, Abdulrazak B. Ibrahim, Benjamin Ewa Ubi

For more information about this series, please visit: https://www.routledge.com/Multidisciplinary-Applications-and-Advances-in-Biotechnology/book-series/MAABNIG

Microbiomes and Emerging Applications

Edited by

Nwadiuto (Diuto) Esiobu,
James Chukwuma Ogbonna,
Charles Oluwaseun Adetunji,
Olawole O. Obembe,
Ifeoma Maureen Ezeonu,
Abdulrazak B. Ibrahim, and
Benjamin Ewa Ubi

CRC Press
Taylor & Francis Group
Boca Raton London New York

CRC Press is an imprint of the
Taylor & Francis Group, an **informa** business

First edition published 2022
by CRC Press
6000 Broken Sound Parkway NW, Suite 300, Boca Raton, FL 33487-2742

and by CRC Press
2 Park Square, Milton Park, Abingdon, Oxon, OX14 4RN

CRC Press is an imprint of Taylor & Francis Group, LLC

Library of Congress Cataloging-in-Publication Data
Names: Ubi, Benjamin Ewa, 1968- editor. | Biotechnology Society of Nigeria, issuing body.
Title: Multidisciplinary applications and advances in biotechnology : contributions from the Biotechnology Society of Nigeria working groups, six-volume set / edited by Benjamin Ewa Ubi [and fourteen others].
Description: First edition. | Boca Raton : CRC Press, 2022. | Includes bibliographical references and index. |
Contents: Volume 1. Fermentation and Algal Biotechnologies for the Food, Beverage and Other Bioproduct Industries — Volume 2. Agricultural biotechnology, biodiversity and bio-resources conservation and utilization — Volume 3. Medical Biotechnology, Biopharmaceutics, Forensic Science and Bioinformatics — Volume 4. Biosafety and Bioethics in Biotechnology — Volume 5. Microbiomes and Emerging Applications — Volume 6. Bioenergy and environmental biotechnology for sustainable development, |
Summary: "This six volume book set examines a range of topics and applications related to biotechnology. Volumes include fermentation and algal biotechnologies; agricultural biotechnology; medical biotechnology, biopharmaceutics; biosafety, bioethics, biotechnology policy; microbiomes; bioenergy and environmental biotechnology for sustainable development"— Provided by publisher.
Identifiers: LCCN 2021047039 (print) | LCCN 2021047040 (ebook) | ISBN 9780367766948 (v. 1 ; hardback) | ISBN 9781032013763 (v. 1 ; paperback) | ISBN 9780367766955 (v. 2 ; hardback) | ISBN 9781032014982 (v. 2 ; paperback) | ISBN 9780367766962 (v. 3 ; hardback) | ISBN 9781032015026 (v. 3 ; paperback) | ISBN 9780367767020 (v. 4 ; hardback) | ISBN 9781032015781 (v. 4 ; paperback) | ISBN 9780367767044 (v. 5 ; hardback) | ISBN 9781032018331 (v. 5 ; paperback) | ISBN 9780367767051 (v. 6 ; hardback) | ISBN 9781032018416 (v. 6 ; paperback) | ISBN 9781003178378
(v. 1 ; ebook) | ISBN 9781003178880 (v. 2 ; ebook) | ISBN 9781003178903
(v. 3 ; ebook) | ISBN 9781003179177 (v. 4 ; ebook) | ISBN 9781003180241
(v. 5 ; ebook) | ISBN 9781003180289 (v. 6 ; ebook)
Subjects: LCSH: Biotechnology—Nigeria.
Classification: LCC TP248.195.N6 M95 2022 (print) | LCC TP248.195.N6
(ebook) | DDC 660.609669—dc23
LC record available at https://lccn.loc.gov/2021047039
LC ebook record available at https://lccn.loc.gov/2021047040

ISBN: 978-0-367-76704-4 (hbk)
ISBN: 978-1-032-01833-1 (pbk)
ISBN: 978-1-003-18024-1 (ebk)

DOI: 10.1201/9781003180241

Typeset in Times
by codeMantra

This book is dedicated to Professor Vincent Ado Tenebe, FBSN, Chairman of the Board of Trustees (BoT) of the Biotechnology Society of Nigeria (BSN) for his passion for Biotechnology and commitment to the cause of the Society.

Contents

Foreword xi
Preface xiii
Editors xv
Contributors xix

Chapter 1 Microbiome and Metagenomics Researches in Nigeria: Evolution, Current Status and Prospects 1

Francisca O. Nwaokorie, Ime R. Udotong, Nwadiuto (Diuto) Esiobu, and Ifeoma Enweani-Nwokelo

Chapter 2 Microbial Biotechnology: Expanding the Frontiers of Microbiomes for Innovation in Agriculture and Medicine 17

Olawole O. Obembe, Nwadiuto (Diuto) Esiobu, O. S. Aworunse, and Nneka R. Agbakoba

Chapter 3 The Need for Probiotics in Nigeria 57

Kingsley C. Anukam and Nneka R. Agbakoba

Chapter 4 Development and Composition of the Human Microbiome from Birth 67

Toochukwu E. Ogbulie, Nwadiuto (Diuto) Esiobu, and Muinah Fowora

Chapter 5 Microbiota Transplantation, Health Implications, and the Way Forward 79

Olugbenga Samuel Michael, Olufemi Idowu Oluranti, Ayomide Michael Oshinjo, Charles Oluwaseun Adetunji, Juliana Bunmi Adetunji, and Nwadiuto (Diuto) Esiobu

Chapter 6 Microbiome Reshaping and Epigenetic Regulation 99

Olugbenga Samuel Michael, Olufemi Idowu Oluranti, Ayomide Michael Oshinjo, Charles Oluwaseun Adetunji, Kehinde Samuel Olaniyi, and Juliana Bunmi Adetunji

Chapter 7 Probiotic Research Studies That Led to the Discovery of *Lactobacillus pentosus* KCA1 121

Kingsley C. Anukam

Chapter 8 A Brief Review of Earth Microbiomes and Applications 137

Toochukwu E. Ogbulie, Nwadiuto (Diuto) Esiobu, and Ifeoma Enweani-Nwokelo

Index 161

Foreword

Microbiome research focuses on the behaviour, interactions and the role of microorganisms living within a specific environment in a community. Interestingly, a lot of progress has been made in microbiome research in the past decade. During this period, an enormous amount of data has been accumulated by researchers via high-throughput assays. This is in part due to the decreasing cost of these assays as well as the reduced cost of data analysis, fuelled by the increase in computing power within this period, made possible by the use of readily available hardware devices and software applications in processing huge amounts of available data in a relatively short time.

This increased effort in microbiome research is driven by recent scientific and technological advances which have helped to unravel the nature of microbial communities, including their interactions with each other and with their environment as well as the resultant effects of such associations within the ecological community. Advancing our current understanding of microbiota and their roles resulting from their unique interactions with other microbes and the environment where they exist has generated enormous interest in the global research community. This is because things learned can be constructively organized into state-of-the-art diagnostic techniques and interventional strategies with enormous applications in diverse fields of study in key disciplines such as human and animal medicine, agriculture and food safety, ecology, diagnostic and forensic science, and pharmaceutical sciences, among others.

One of the major goals of this book which I consider very timely is to discuss multidisciplinary opportunities and new developments in biotechnology with reference to microbiomes and their emerging applications. The book starts off with an overview of microbiome and metagenomic research in Nigeria. In the ensuing chapters, it introduces the reader to the emerging challenges, exciting possibilities, and current trends in microbiome research and other potential applications.

The book is broken into eight key topic areas: (1) Microbiome and metagenomics researches in Nigeria: evolution, current status and prospects; (2) The need for probiotics in Nigeria; (3) Development and composition of the human microbiome from birth; (4) Microbiome reshaping and epigenetic regulation; (5) A brief review of earth microbiomes and applications; (6) Probiotic research studies that led to the discovery of *Lactobacillus pentosus* KCA1; (7) Microbial biotechnology: expanding the frontiers of microbiomes for innovations in agriculture and medicine; and (8) Microbiota transplantation, health implications and the way forward.

As such, this book is a huge resource for both teachers and students in higher institutions. It not only explores available opportunities and priorities but also provides current directions for future investigation and multidisciplinary collaboration. I recommend this book to teachers and students especially those involved in this and related subject areas.

Emeka I Nweze, MSc, AIMLS, PhD, MPH
Professor,
Department of Microbiology,
University of Nigeria, Nsukka

Preface

In recognition of the pivotal role of biotechnology as a burgeoning technology in fostering a knowledge-driven bio-based economy, the Biotechnology Society of Nigeria (BSN) has – in the about 40 years of its founding – made enormous contributions towards expanding the frontiers of biotechnology in Nigeria for sustainable livelihoods. This book is one of the multi-volume series on the Multidisciplinary Applications and Advances in Biotechnology: Contributions from the Biotechnology Society of Nigeria (BSN), with each volume covering specific contemporary topics targeted at providing scientists in the academia, industry, research institutes/agencies, etc. with recent advances in the field so as to bridge the knowledge gap in the crucial areas of biotechnology.

Designed to document advances in biotechnology in Nigeria over the past decades and the emerging biotech sector as a key driver of the bioeconomy, this book is a timely knowledge product. It documents key issues on recent advances in the field for use by a variety of readers including postgraduate students, professionals in the field, industry players, policy makers, science advocacy groups, etc. and engenders a deeper understanding of underlying mechanisms, technologies, processes and science-policy nexus that have placed Nigeria as a leader in biotechnology in Africa.

This volume V of this book series on microbiomes and emerging technologies emanates from the Microbiomes and Microbial Biotechnology Working Group of the Society (WG14). It contains a total of eight chapters that bring to light the applications of available knowledge and deliverables in the rapidly expanding area of microbial biotechnology covered in the topics by the array of seasoned contributing authors. The chapters cover a wide range of topics including the evolution of microbiomes and metagenomics research in Nigeria, its current status and prospects, and expanding the microbiomes' frontiers for innovations in agriculture and medicine. It also covers topics such as the growing need for probiotics in Nigeria, the human microbiome and development from birth, and microbiota transplantation and health implications. Other topics include microbiome reshaping and epigenetics; probiotic research studies in Nigeria and Canada that led to the discovery of *Lactobacillus pentosus* KCA1; and insights on the earth's microbiomes and their applications for economic development and sustainable ecosystems. Each chapter addresses the relevant research area, discusses each of the thematic areas concisely, emphasizes the important points and largely illustrates research processes beautifully with clear schematic diagrams to ease comprehension of the presented subject areas. The chapters are co-authored by a team of different scientific experts in the field, thereby producing in-depth scholarly output in a high-level of interdisciplinary collaborative manner, using the pool of expertise available in the Biotechnology Society of Nigeria (BSN) Microbiomes and Microbial Biotechnology Working Group. As such, it would go a long way in ensuring a higher level of sustainability on the deliverables expected from these thematic research areas towards fostering a sustainable knowledge-driven bio-based economy.

This book is a useful reference material, a comprehensive description of the recent advances in the field of microbiomes and microbial biotechnology and a helpful guide to the subject matter's original literature. It is specifically targeted at scientists and researchers working microbial biotechnology, probiotics, nature health, biopharmaceuticals, nutraceuticals, fermentation technologies and industrial biotechnology, environmental sustainability, etc. in the universities, research institutions, industry or government agencies.

Overall, this volume is an excellent piece of scientific contribution that targets a wide range of audience including global leaders working on strategies at combating malnutrition, particularly in sub-Saharan Africa, industrialists and biotech start-ups, the agriculture and food industry, the health sector, professionals involved in biopharmaceutics and therapeutics, investors, farmers, policy makers, educators and researchers, and educational and research institutions (most especially for postgraduate students). Readers will find this book a valuable resource containing recent advances contributed by a seasoned and diverse array of scholars within this professional society.

The carefully selected editorial team and the authors of individual chapters drawn from several key institutions nationally and internationally were chosen for their recognized global expertise and contributions to the various subject areas. Their willingness to make these invaluable contributions in their respective fields of expertise in the context of this Working Group and in a collaborative way to foster research networking is gratefully acknowledged. The editors wish to thank the different manuscripts' reviewers for their unceasing dedication and great input during the peer-review process. We thank Prof. Malachy O. Akoroda of the Department of Agronomy, University of Ibadan, who shared the initial idea that inspired this book initiative with me just after the conclusion of the 32nd Annual Conference of the Society held at the International Institute of Tropical Agriculture (IITA) Ibadan, Nigeria. Furthermore, we highly appreciate the foresight, diligent support and understanding of the publisher, Taylor and Francis Group (USA), for publishing *Multidisciplinary Applications and Advances in Biotechnology – V: Microbiomes and Emerging Applications* with their international hallmark of excellence. We return special thanks to Dr. Marc Gutierrez and Mr. Nick Mould, whose great and painstaking efforts, constant suggestions and guidance at all stages saw to the successful publication of this book.

Prof. Benjamin Ewa Ubi, Ph.D.
President, Biotechnology Society of Nigeria (BSN)
& Chief Editor
13 February, 2021

Editors

Nwadiuto (Diuto) Esiobu (Ph.D., University of Louvain, Belgium; Post Doc., Massachusetts Institute of Technology, Cambridge, USA) is currently a Professor of Microbiology and Biotechnology at Florida Atlantic University, Boca Raton, FL, USA, where her microbial biotech laboratory integrates environmental genomics and health risk assessment with rapid pathogen detection using synthetic biology. Her research interests in agriculture and health include molecular ecology of plant invasion, soil fertility and the applications of microbiomes in humans and plants. She has served in senior professional leadership roles in academia, industry and policy, including selection panel for the US President's award for excellence in Math and Science (Washington, DC) and more than 11 years on Committees of the Board of the American Society for Microbiology. A Fellow of the Biotechnology Society of Nigeria and a Jefferson Science Fellow, she was selected by the National Academy of Sciences as a senior science advisor to Secretary Clinton's office of Global Food Security (GFS) and the Office of International Health and Biodefense (OES/IHB), US Department of State, Washington, DC, and named world-class Faculty by the City of Fort Lauderdale for her contributions to the scientific enterprise.

James Chukwuma Ogbonna, Ph.D., is a Professor of Industrial Microbiology and Biotechnology. He obtained a B.Sc. (First-Class Hons in Botany) from University of Jos, Nigeria (1982); M.Eng. (Fermentation Technology) from Yamanashi University, Japan (1988); and a Ph.D. (Applied Biochemistry) from University of Tsukuba, Japan (1991). He is currently the Director of National Biotechnology Development Agency, South East Zonal Biotechnology Centre, University of Nigeria, Nsukka, and the immediate past Deputy Vice Chancellor (Academic), University of Nigeria Nsukka. He was the National President, Biotechnology Society of Nigeria (2010–2014), and the current President of the Foundation for African Development through International Biotechnology. His current research interests include photobioreactor design and optimization, development of microbial cell culture systems and applications and bio-energy production. He has published 3 books, co-edited 3 books, published 13 book chapters and 111 peer-reviewed journal articles, and presented 9 conference proceedings and over 120 papers in national and international conferences in over 11 countries. His current H-index is 34 with over 3,760 citations in Google Scholars. Prof. Ogbonna is a Fellow of the Biotechnology Society of Nigeria, Fellow of the Foundation for African Development through International Biotechnology and Fellow of the Institute of Corporative Administration.

Olawole O. Obembe is a Professor and the UNESCO Chair-holder on Plant Biotechnology at Covenant University, Nigeria. He earned a Ph.D. in Plant Science from Wageningen University, the Netherlands, while his bachelor's and Master's degrees were obtained from the Obafemi Awolowo University, Ile-Ife, Nigeria. He was a Research Fellow at the International Institute of Tropical Agriculture (IITA)

Ibadan from 1998 to 2001. He was also a WOTRO and Wageningen University Ph.D. Scholar from 2001 to 2006. Moreover, he was a postdoctoral fellow of the UNESCO's International Centre for Genetic Engineering and Biotechnology, Trieste, Italy (2009-2010). He was appointed Senior Lecturer at Covenant University, Nigeria, in 2007 and rose to a full professor position in 2014. He has held various administrative positions in the university, such as Deputy Dean of College, Director of Industry Partnership Unit, the Centre for Research, Innovation and Development, and the Biotechnology Centre. He is currently the Chair-holder for the UNESCO Chair on Plant Biotechnology in Covenant University. His research interest includes genetic diversity and genomic studies in indigenous Nigerian plants; *in vitro* conservation, regeneration and multiplication of endangered (medicinal) plants; plant genetic engineering; production of biopharmaceutics and industrial proteins in plants; and identification and isolation of lead compounds in plants for new drugs discovery. He has published over 70 peer-reviewed articles in reputable journals and conference proceedings.

Prof. Obembe has served on the editorial board of numerous international journals, including Frontiers in Plant Science, International Journal of Biotechnology Applications, Biosafety Journal, Journal of Virology and Biotechnology, Indian Journal of Science and Technology, Journal of Microbial and Biochemical Technology, and Asian Pacific Journal of Tropical Medicine. He is also a reviewer to many journals, including Plant Cell Reports, Molecular Biology Reports, Journal of Ethnobiology and Ethnomedicine, African Journal of Biotechnology, Journal of Plant Growth Regulation, Biotechnology and Molecular Biology, Nigerian Journal of Biotechnology. He is a member of several professional bodies, including the Biotechnology Society of Nigeria, Botanical Society of Nigeria, African Biotechnology Network, and the Society for Applied Biotechnology, India.

Professor Ifeoma Maureen Ezeonu, B.Sc. (Nig.), M.Sc., Ph.D. (Atlanta), is a Professor of Medical Microbiology and Molecular Genetics in the Department of Microbiology, University of Nigeria, Nsukka, and currently the Director of the University of Nigeria Central Science Laboratory. She obtained her B.Sc. in Microbiology from the University of Nigeria, Nsukka in 1987 and her M.Sc. and Ph.D. in Microbiology from Georgia State University, Atlanta, Georgia, USA in 1991 and 1994, respectively. Her areas of research include bacteriology, medical biotechnology, molecular biology, bacterial drug resistance, probiotics and prebiotics. She is the Coordinator of the Drug-Resistant Infections Research Group of the University of Nigeria.

Charles Oluwaseun Adetunji, Ph.D., is an Associate Professor of Microbiology and Biotechnology at the Microbiology Department, Faculty of Science, Edo State University Uzairue EDSU, Edo State, Nigeria, where he utilized the application of biological techniques and microbial bioprocesses for the actualization of sustainable development goals and Agrarian revolution, through quality teaching, research and community service. He is currently the Acting Director of Intellectual Property and Technology Transfer, Sub-Dean for Faculty of Science, and the Head of department of Microbiology at EDSU. He is a Visiting Professor and the Executive Director for the Center of Biotechnology, Precious Cornerstone University, Ibadan.

He has won several scientific awards and grants from renowned academic bodies like Council of Scientific and Industrial Research (CSIR), India; Department of Biotechnology (DBT), India; the World Academy of Science (TWAS), Italy; Netherlands Fellowship Programme (NPF), the Netherlands; the Agency for International Development Cooperation; Israel; and Royal Academy of Engineering, UK, among many others. He has filed several scientific patents on Bioherbicides, Biopesticides, Nanobiosurfactants, Nanobiopesticdes and many more. He has published over 300 scientific articles in refereed journals and conference proceedings both international and local journals. He was ranked among the top 500 prolific authors in Nigeria between 2019 till date by SciVal/SCOPUS. He is currently editing several biotechnology text books with Elsevier. He was recently appointed as the President and Chairman Governing Council of the Nigerian Bioinformatics and Genomics Network Society. He is presently a series editor with Taylor and Francis, USA editing several textbooks on Agricultural Biotechnology, Nanotechnology, Pharmafoods, and Environmental Sciences. The breadth of his scholarly contributions to research is evident from his contributions which cover several topics relating to Food Security, Agriculture and Environmental Sustainability. His research interest includes microbiology, biotechnology, bioresources and nanotechnology. He is an editorial board member of many international journals and serves as a reviewer to many double-blind peer-review journals like Elsevier, Springer, Taylor and Francis, Wiley, PLOS One, Frontiers, Nature, American Chemistry Society, Bentham Science Publishers, etc. He is a member of many scientific and professional bodies including the American Society for Microbiology, Nigerian Young Academy, Biotechnology Society of Nigeria and Nigerian Society for Microbiology. He has won a lot of international recognition and also acted as a keynote speaker delivering invited talks/position papers at various universities, research institutes and several centres of excellence which span across several continents of the globe. Over the last 15 years, he has built strong working collaborations with reputable research groups in numerous and leading universities across the globe. He is the convener for Recent Advances in Biotechnology, which is an annual international conference where renowned Microbiologists and Biotechnologists come together to share their latest discoveries (https://raibconference.wixsite.com/website). He is the Founder & CEO of BECTIK Biotechnology and Nanotechnology Company. His Biotechnology Company consists of a team of leading academics in their fields and thrives to deliver solutions all around Bio- and Nanotechnology. The company also delivers solutions that are tailored to your business needs and for any scale.

Abdulrazak B. Ibrahim, Ph.D., is an agricultural biotechnologist with experience and skills on plant biotechnology stewardship, stakeholder engagement, foresight analysis, development of transgenic crops, laboratory and field management of GM crops, biosafety analysis, adoption and discontinuation of events. He was trained in the Brazilian Agricultural Research Cooperation (EMBRAPA Center for Genetic Resources), from where he obtained a doctorate degree in Molecular Biology from Universidade de Brasilia (UnB) and M.Sc. from Universidade Federal do Ceara (UFC), Fortaleza, and acquired best practices in agricultural biotechnology leading to the development and patenting of a transgenic technology for the control of whitefly using RNA interference technology within EMBRAPA, and the development of virus-resistant cowpea, under the Africa-Brazil Innovation Platform, implemented by EMBRAPA and Forum for Agricultural Research in Africa (FARA), with support from the Department for International Development (DFID), Bill & Melinda Gates Foundation (BMGF), International Fund For Agricultural Development (IFAD) and Foundation Arthur Bernardes (FUNARBE).

Benjamin Ewa Ubi, Ph.D., is a Professor of Plant Breeding and Biotechnology. He obtained his B.Sc. Honours (Agric.) Agronomy from the University of Calabar (1990) and M.Sc. Agronomy – Crop Science (Plant Breeding and Tissue Culture) (1994) and Ph.D. Agronomy (Plant Breeding and Biotechnology) (1998) from the University of Ibadan, Nigeria. His Ph.D. research was completed at the then Biotechnology Laboratory of the International Institute of Tropical Agriculture (IITA), Ibadan, where he was a graduate research fellow and became the President of the International Association of Research Scholars and Fellows (IARSAF) in 1996. Among several other research engagements, Prof. Ubi served as Science and Technology Agency (STA) Postdoctoral Fellow under the Japan International Science and Technology Exchange Centre (JISTEC) 1999–2001 at the Plant Breeding Department (Biotechnology Laboratory) of the National Institute for Livestock and Grassland Science (NARO, National Agricultural Research Organization), Nishinasuno, Japan; and as a three-term Visiting Scientist at the NARO National Institute of Fruit Tree Science (NIFTS), Tsukuba, Japan, under the auspices of the Inoue Research Foundation of Japan Fellowship (April 2003–April 2004) and the Japanese Society for the Promotion of Science (JSPS) visiting fellowships (October 2008–October 2009; May 2012–March 2013). Prof. Ubi served as Visiting Professor at the Arid Land Research Centre (The Global Centre of Excellence for Dryland Science) (Molecular Breeding Laboratory), Tottori University, Japan; and Visiting Professor University of Tokyo, Japan. Among several other academic engagements, Prof. Ubi has served as the Director of the Biotechnology Research and Development Centre, Ebonyi State University, Abakaliki (August 2006–March 2008) and the Head of the Department of Biotechnology, Ebonyi State University for several years; Dean, Faculty of Science, Ebonyi State University Abakaliki and Pioneer Editor-in-Chief of *Science Journal*, https://sciencejournal.ng, a publication of the Faculty of Science, Ebonyi State University, Abakaliki; President, Biotechnology Society of Nigeria (BSN) and Leader of the Crops Genetics, Breeding and Genetics Working Group of the Biotechnology Society of Nigeria (BSN).

Contributors

Charles Oluwaseun Adetunji
Microbiology, Biotechnology and Nanotechnology Laboratory
Department of Microbiology
Edo State University Uzairue
Edo State, Nigeria

Juliana Bunmi Adetunji
Nutritional and toxicological Research Laboratory
Department of Biochemistry Sciences
Osun State University
Osogbo, Nigeria

Nneka R. Agbakoba
Department of Medical Laboratory Science
Nnamdi Azikiwe University
Awka, Nigeria

Kingsley C. Anukam
Faculty of Health Sciences & Technology
Department of Medical Laboratory Science
Nnamdi Azikiwe University
Awka, Nigeria
and
Uzobiogene Gnomics
Ontario, Canada

O. S. Aworunse
Department of Biological Sciences
Covenant University
Ota, Nigeria

Ifeoma Enweani-Nwokelo
Department of Medical Laboratory Science
Nnamdi Azikiwe University
Awka, Nigeria

Nwadiuto (Diuto) Esiobu
Applied Biotech International Nigeria Ltd
Wuye, Nigeria
and
Microbial Biotechnology Laboratory
Biology Department
Florida Atlantic University
Boca Raton, USA

Muinah Foworah
Molecular Biology Unit
Nigerian Institute of Medical Research (NIMR)
Yaba Lagos, Nigeria

Olugbenga Samuel Michael
Cardiometabolic Research Unit
Department of Physiology, College of Health Sciences
Bowen University
Iwo, Nigeria

Francisca O. Nwaokorie
Department of Medical Laboratory Science
College of Medicine
University of Lagos
Lagos State

Olawole O. Obembe
UNESCO Chair on Plant Biotechnology
Plant Science Research Cluster
Department of Biological Science
Covenant University
Ota, Japan

Toochukwu E. Ogbulie
Department of Biotechnology
School of Biological Sciences
Federal University of Technology
Owerri, Nigeria

Kehinde Samuel Olaniyi
Afe Babalola University
Ado Ekiti, Nigeria

Olufemi Idowu Oluranti
Environmental Physiology Research Unit
Department of Physiology, College of Health Sciences
Bowen University
Iwo, Nigeria

Ayomide Michael Oshinjo
Cardiometabolic Research Unit
Department of Physiology, College of Health Sciences
Bowen University
Iwo, Nigeria

Ime R. Udotong
Department of Microbiology
University of Uyo
Uyo, Nigeria

1 Microbiome and Metagenomics Researches in Nigeria

Evolution, Current Status and Prospects

Francisca O. Nwaokorie
University of Lagos

Ime R. Udotong
University of Uyo

Nwadiuto (Diuto) Esiobu
Applied Biotech International Nigeria Ltd
Florida Atlantic University

Ifeoma Enweani-Nwokelo
Nnamdi Azikiwe University

CONTENTS

1.1 Introduction 2
1.2 Microbiome 3
1.3 Metagenomics 3
1.4 Application of Metagenomics 3
1.5 Products from Microbiomes and Metagenomics 4
1.6 History and Evolution of Microbiomes and Metagenomics Researches in Nigeria 4
1.7 Metagenomic Sequencing Approaches for Investigating Animal, Human and Soil Microbes in Nigeria 6
1.8 Status of Metagenomics Services in Nigeria 6
1.9 Current Academic, Health and Industrial Structures Supporting Microbiome and Metagenomics Research in Nigeria 8
1.10 Challenges of Microbiome and Metagenomics Research in Nigeria 8
1.11 Prospects of Human Microbiome and Metagenomics Research 11

DOI: 10.1201/9781003180241-1

1.12 Recommendations 12
1.13 Conclusion 13
References 13

1.1 INTRODUCTION

Microbiomes and metagenomics researches have come to revolutionize microbial ecology in particular and microbiological researches in general. Though it is as old as over three decades, it is still considered new and novel in Nigeria. Till date, many are unaware of its vast applications in medicine, agriculture, food, biotechnology, environmental remediation and sustainability and indeed all spheres of human endeavours, and the importance aspiration that it holds for the future. Following trends, the quantum of research on human and environmental microbiomes and metagenomics researches is underscored by very little data emanating from Nigeria. Globally, thousands of microbiome research projects have been carried out by various researchers.

Metagenomic analyses are applied in the fields of agriculture, biofuel, biotechnology, ecology, environmental remediation, and pharmaceutical and medical sciences (human microbiome characterization and infectious disease diagnosis). Important reviews and presentations of some seminal microbiome projects include the human microbiome (Gevers et al., 2012), the earth microbiome (Thompson et al., 2017), ocean microbiome (Zehr et al., 2017), foetal microbiome (Ma et al., 2014), newborn microbiome (Groer et al., 2014), and cadaver microbiome (Gilbert and Lynch, 2019), among similar projects in various countries. The benefits of the research emanating from this field consist of products available in human microbiome market. This includes prebiotics, probiotics, foods, and medical diagnostic tests and drugs. In addition, this technology has wide application in therapeutics and management of infectious diseases, metabolic/endocrine disorders, inflammatory and cardiovascular disease, cancer, blood and neurological disorders. Moreover, there are products for research technology in use for sequencing, genomics, proteomics, metabolomics and other 'OMICS.

By using next-generation sequencing (NGS) technologies in Nigeria, some researchers have successfully brought to light the actual composition of microorganisms in different environments in Nigeria (Olukunle and Boboy, 2013; Udotong et al., 2015; Ogbulie and Nwaokorie, 2016; Udotong et al., 2017; Ezekoye et al., 2018; Uko et al., 2019, Anika et al., 2020). Microbiomes and metagenomic studies are pivotal to unlocking the agricultural potential of Nigeria in a renewable, climate-smart and sustainable approach. Understanding the microbiomes of Nigerian soils and how these could be manipulated for prime fertility and soil quality has also been addressed through these types of studies (Bukar et al., 2019). Studies in the area of basic medical sciences including that on vagina, gut and oral microbiome also exist (Ayeni et al., 2018; Anukam and Agbakoba, 2017; Anukam et al., 2019; Nwaokorie et al., 2020).

Most of Nigerian researchers acquired the skills through hands-on training on DNA extraction and standardization in local institutions. However, the core metagenomic sequencing experiences are obtained in foreign laboratories through collaborative research, personal sponsorship, fellowships and grants. Looking at institutional capacity, African Centre of Excellence for Genomics of Infectious Diseases (ACEGID) at Redeemer's University, Ede, is drawing synergistically on the intersection between training, research and health (Dudas et al., 2017; Ajogbasile et al., 2020). The African Centre of Excellence for Phytomedicine Research and Development (ACEPRD), University of Jos, functions under similar mandate. Irrespective of the contributions made so far, accessibility to these centres remains limited as most samples still find their ways outside Nigeria for NGS analysis showing that the available infrastructure is far below the national needs required to facilitate metagenomics researches. Till date, outputs recorded have not translated these basic researches to patent, industrialization and commercialization for national development.

1.2 MICROBIOME

Microbiome refers to the collection of genomes of microorganisms (microbiota) living in a particular environment. Virtually everything in nature, both living and non-living things, harbour microorganisms and can therefore be referred to as "microbiomes". Microbiomes is used to describe communities of various types of microbes, their gene and gene products collectively in a selected habitat be it on humans, animals, vegetation, soils, oceans or even buildings (Blaser et al., 2016). Microbiomes describe the well-studied bacteria consortia, while virobiomes and mycobiomes apply to viruses and fungi, respectively. The scientific concept of metagenomic approach evolved in an effort to discover more about the diversity of these microbes in natural environments (López-García and Moreira, 2008).

1.3 METAGENOMICS

Metagenomics is a molecular biology tool that involves simultaneous sequencing and analysis of a wide range of genomes in a niche, including functional analysis following gene annotation. The discipline involves the use of culture-independent genomic analysis to provide information on the microbial diversity and ecology of a specific environment (Handelsman et al., 1998). Taxonomic profiling, characterization and analysis of these microbial communities are performed using NGS platforms. The most common platforms are 454 Life Sciences (Roche) and Illumina systems (Solexa) (Oulas et al., 2015). Others include SOLiD 5500 W Series (Applied Biosystems), single-molecule real-time (SMRT) DNA sequencing (Pacific Biosciences), Ion Torrent semiconductor sequencing and Strand sequencing technologies (Oxford Nanopore technologies) as well as Irys Technology (BioNano Genomics) (Oulas et al., 2015; Jain et al., 2016).

1.4 APPLICATION OF METAGENOMICS

The most widely employed sequencing technology is the amplicon metagenomics where sub-genomes or genes (partial or entire) are studied using universal primers. Whole genome metagenomics reveals the entire microbial makeup of a niche and is often used in public health risk assessment food safety and in discovering and mining bioresources (Mercer et al., 2020). The tools for metagenomic analysis are the DNA processed through workflow of library preparation, sequencing and bioinformatics analysis (Figure 1.1).

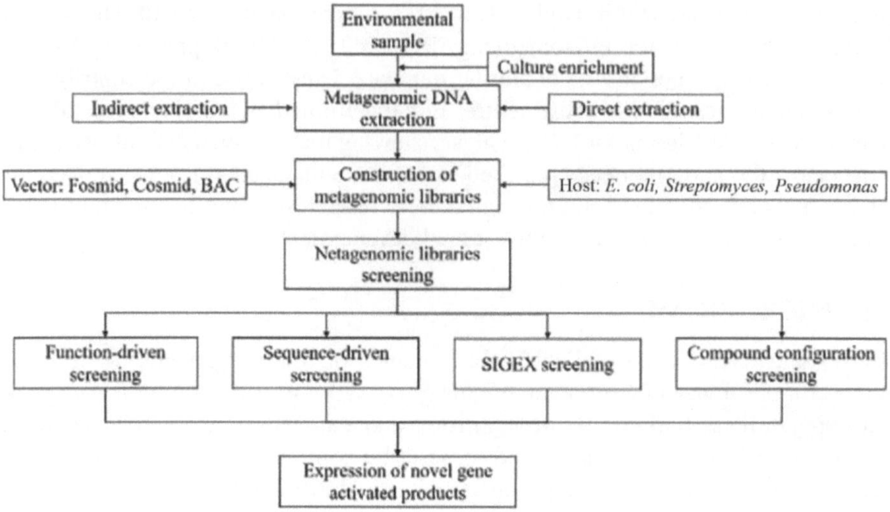

FIGURE 1.1 General process of metagenomic workflow. (Source: Awasthi et al., 2020)

Metagenomic analyses are applied in the field of agriculture, biofuel, biotechnology ecology, environmental remediation, pharmaceutical and medical sciences (human microbiome characterization and infectious disease diagnosis). The technology has been used extensively to define microbial diversity, bioremediation, pollution monitoring, enzyme and drug discovery. Its applications entail the identification of novel bacteria or gene clusters encoding for enzymes that degrade substrates. In addition, this technology is very useful in the diagnosis and monitoring of pathogens (Datta et al., 2020).

1.5 PRODUCTS FROM MICROBIOMES AND METAGENOMICS

Available products in human microbiome market include prebiotics, probiotics, foods, and medical diagnostic tests and drugs. They are in use for therapeutics and management of infectious diseases, metabolic/endocrine disorders, inflammatory and cardiovascular disease, cancer, blood and neurological disorders. Similarly, in soil metagenomics, activity-based screening has the potential to detect entirely new genes encoding for novel types and classes of enzymes. Such an approach is also used to identify new bioactive compounds and functional gene products. To support this, cellulases, lipases and xylanase enzymes and antibiotics with appreciable commercial values have been isolated from various natural environments like soil, rumen, compost soil and plants, among others (Shrinivas et. al., 2019; Amao et al., 2019).

1.6 HISTORY AND EVOLUTION OF MICROBIOMES AND METAGENOMICS RESEARCHES IN NIGERIA

Embracing microbiome and metagenomic research in Nigeria was initially challenged by lack of skills, project design, funding and opportunity to collaboration with experts in these fields. This was somehow bridged by the involvement of researchers in-country. Around 2010, Adelekan and Esiobu (2012) employed amplicon metagenomics to analyse a local solid-state fermentation product, with the aim of improving its quality and extending its shelf life. Udotong et al. (2015) evaluated the microbial diversity of a remote aviation fuel contaminated sediment of a lentic ecosystem in Ibeno, Nigeria. This study was followed by the determination of procaryotic diversity of the remote aviation fuel-polluted lentic ecosystem in Ibeno, Nigeria (Udotong et al., 2017). They later worked on homology-dependent metagenomics study of agricultural soils in Akwa Ibom State, Nigeria (Uko et al., 2019). Other metagenomic studies focusing on environmental microbiome include the study on microbial community diversity and genetic functions of environmental microbes capable of degrading relevant substrates (Ogbulie and Nwaokorie, 2016); metagenomic insights into effects of spent engine oil perturbation on the microbial community composition and function in a tropical agricultural soil (Salam et al., 2017); and metagenomics of bioslurry obtained from a biogas plant and anaerobic digester using next-generation DNA sequencing (Akubuenyi et al., 2019; Anika et al., 2020). Other studies in that line looked at the correlation and distribution of xenobiotic genes and metabolic activities with the level of total petroleum hydrocarbon in soil, sediment and estuary water in the Niger Delta (Edet et al., 2018a), and evaluation of microbial diversity of epipelic and benthic sediment using cultural and metagenomics techniques (Edet et al., 2018b) as well as that on the potential benefits of applying "OMICS" technology in cleaning up crude oil spillages in the Niger Delta region of Nigeria (Udofia et al., 2018).

The trend in the area of human microbiome is not left out. Indigenous researchers provided earlier information on metagenomic assessment of antibiotic resistance genes from an ecosystem (Edet et al., 2017a, b) and implication of human microbiome diversity in health and disease (Mboto et al., 2018). Other researches like the works of Anukam and Agbakoba (2017) and Anukam et al. (2019) in the area of oral and vaginal microbiomes are well documented. In addition, Nwaokorie et al. (2020) highlighted that the urban regions of Lagos seem to have transitioned towards a diet pattern typical of "Westernized" societies, and this may have contributed to the variation in the microbiome of those of similar ethnicity, age and sex living in the rural region. These few pieces of literature need to be followed up by future studies that would not only define biodiversity and genetic functionality, but also provide indebt information on how this science can be employed in disease modelling, prediction, diagnosis, prevention and treatment.

Environmental microbes have been used extensively in biotechnology. Microbial metabolites are used to design analytical reagents and chemicals including enzymes, antibiotics and other therapeutic products used in the present-day laboratory and personalized medicine. It is envisaged that when functional genes are identified, they can be put into large-scale use in product design and manufacturing, an aspect lacking in Nigeria's research output. Following that trend, few metagenomic researches can be credited to Nigerian researchers. Works published so far are focused basically on the comparative metagenomics and functional profiling of soils microbes in communities with varying pollution history (Chikere et al., 2011, 2019) and monitoring of microbial hydrocarbon remediation in the soil (Chikere et al., 2011; Ezekoye et al., 2018). In addition, Salam and co-workers in 2017 conducted a research into community composition and functional genes of microbes present in agricultural soil polluted with spent engine oil (Salam et al., 2017) and hydrocarbon-contaminated tropical African soil (Salam et al., 2018). Most recently, Fasesan et al. (2020) used metabarcoding techniques to assay the prevalence of novel bacteria with potential use in biotechnology in the Ikogosi warm springs of Nigeria. Similarly, in the field of agriculture, a study profiled the diversity and community structure of bacteria associated with post-harvest Irvingia fruits (Etebu et al., 2018). However, there are paucity of records on how these basic researches may have translated to patent, industrialization and commercialization.

1.7 METAGENOMIC SEQUENCING APPROACHES FOR INVESTIGATING ANIMAL, HUMAN AND SOIL MICROBES IN NIGERIA

The intense number of microbes in an environment requires a technique that captures the genes present. Metagenomics moves microbial research away from the cultural techniques that have been used to study microbes in Nigeria. Metagenomics is a culture-independent technique. It is not indigenous to Nigeria. The protocols and services are acquired and used for research into the soil, human and animal microbiomes. Earlier in 2013, researchers keyed into the application of these techniques following knowledge attained from workshops, seminars and scientific publications. At that time, metagenomic services were done in laboratories outside Nigeria. Those who then had basic skills on molecular techniques took up the challenge to advance knowledge on metagenomics and collaborated with biotechnology laboratories outside Nigeria. Basically, this was made by replicating procedures to extract DNA at almost all the universities and research institutes. Subsequently, standardized DNA was sent to foreign laboratories for metagenomic sequencing and this is expected to follow national ethical guidelines for transfer of DNA.

Metagenomic analysis was first carried out using Roche/454 platform. Bioinformatics information following sequencing was provided on BLAST hit, Internal Label, frequency of BLAST hit in the samples as well as count reads corresponding to the BLAST hit. From 2015, similar Nigerian samples were analysed following the introduction of Illumina MiSeq Platform. This technique was used in the analysis of microbial diversity of a remote aviation fuel contaminated sediment of a lentic ecosystem in Ibeno, Nigeria (Udotong et al., 2015). Many other studies followed in that line, and commercialized metagenomic services were obtained from different countries abroad to study soil samples, sediments, shipwrecks, anaerobic digesters, animals (chicken dungs), fruits and human faeces, and oral and vaginal microbiomes. Bioinformatic analysis thereafter expanded to the use of more platforms like Quantitative Insight into Microbial Ecology (QIIME) (Kuczynski et al., 2012) and Kyoto Encyclopedia of Genes and Genomes (KEGG) (Kanehisa and Goto, 2000).

1.8 STATUS OF METAGENOMICS SERVICES IN NIGERIA

There are a reasonable number of human resources available in Nigeria. With a lack of relevant infrastructure researches, university lecturers and postgraduate students have acquired skills for metagenomic sequencing through hands-on workshops organized and supported by Wellcome Trust, UK, Keystone Symposia, OMICS International pre-conference and abstracts presented at conferences and stepped down across Nigeria. There is expertise in metagenomic techniques, and there are researchers in universities, research institutions and biotechnology companies. A good number of these resource persons are involved in collaborative research locally and internationally. They also engage themselves in supervisory activities to guide the research to ensure the data can address the researchers' questions. This service is seen more or less to support research conducted by postgraduate students. Unfortunately, research designs are not directed towards short- and long-term strategic plans aimed to address national problems. As such, there are pockets of unharmonized research on topics that neither address local needs nor translate to national development.

Similarly, resources available are primarily focused on standardization of genomic extractions and libraries before sending to countries like South Africa, the USA, Germany and South Korea for NGS. Infrastructure is relatively poor and limited to services that would address pre-sequencing processes. Similarly, few experts are available for bioinformatics analysis, and unfortunately, their services are not centralized. There is no existing national database hosting genomic information from outputs of huge genomic data generated in-country or in association with foreign collaborators. Amidst the paucity of infrastructure (Illumina MiSeq, HiSeq, Pac Bio Sequencing machine) for core sequences, genomic extractions are basically available in almost every higher institution in Nigeria either through a central facility or through individual researcher.

Since 2013, metagenomics services have been conducted in laboratories not situated in Nigeria, and till date, that has not changed. DNA samples from animals, humans and soil, among others, are sent out of the country. This poses serious challenges to ethical, economic and health issues. Internal activities consist of experienced indigenous researcher passing on knowledge and skills acquired through hands-on workshops, OMICS International pre-conference, abstracts presented at conferences held across Nigeria and subsequent publications in peer-reviewed journals. Unfortunately, this does not translate to the enormous benefits of the microbiome and metagenomics approaches and services.

Currently, information on the availability and location of NGS machines are very minimal. Looking at institutional capacity, ACEGID at Redeemer's University, Ede, established in 2013 with support and funding from the World Bank, is one of the genomic reference and research centres in Nigeria. This facility is drawing synergistically on the intersection between training, research and health. The reports that viral genomes reveal factors that spread and sustained the Ebola epidemic (Dudas et al., 2017), Real-time metagenomic analysis of undiagnosed fever cases unveils a yellow fever outbreak in Edo State, Nigeria and field evaluation of a Pan-Lassa rapid diagnostic test during the 2018 Nigerian Lassa fever outbreak (Ajogbasile et al., 2020) are credited to researchers from this facility. Similar service is available at the ACEPRD, University of Jos. Looking at the number of research institutions, universities and allied organization in Nigeria, this low number of project-based NGS laboratories is not enough to meet up demands and potentiate benefits of this technology.

1.9 CURRENT ACADEMIC, HEALTH AND INDUSTRIAL STRUCTURES SUPPORTING MICROBIOME AND METAGENOMICS RESEARCH IN NIGERIA

There are ongoing efforts in academic institutions to generate data with little structure to support research. That notwithstanding, there are few facilities in various organizations across the country where some aspects of molecular studies are undertaken. In academic-based institutions, courses are offered at postgraduate levels on microbiome and metagenomics studies with some organizations installing equipment to undertake nucleic acid extraction/amplification. Once this is achieved, the nucleic acid is sent overseas for sequencing. Most researchers in Nigeria had engaged the services of companies in South Africa, the USA and the UK.

Various Nigerian samples have been subjected to microbiome and metagenomics studies. Basically, they are clinical and environmental samples. Clinical samples are derived from but not limited to human and animal anatomical sites: gut, oral cavity, skin and vagina. These samples include stool, urine, blood, skin, saliva, semen and a high vaginal swab. Environmental samples include water, soils, sediments and effluents. Anuka and Agbakoba (2017), reported oral diversity and bacterial community of three Nigerian females of different age groups. Sequencing of the extracted DNA from the oral samples was carried out using Illumina MiSeq Platform. They also conducted a similar study on seminal fluids (Ndiokwere et al., 2019) and evaluated the comparative abundance and functional biomarkers of the vaginal microbiome of women with bacterial vaginosis (Anukam et al., 2019). There are also ongoing studies in Nnewi on the microbiome and metagenomics of Nigerian children's guts with or without diarrhoea episodes. Preliminary results had shown that gut microbiome of Nigerian children consisted of 21 bacterial phyla, the most prevalent being Firmicutes, Bacteroides and Proteobacteria in that order (Unpublished data). Other ongoing researches include those of skin, guts and urine of diabetic patients and in Jos, virginal microbiome and metagenomics in relation to cancer.

Amao et al. (2019) studied the bacterial diversity of the heaps of cassava peels to screen for cellulase and xylanase enzyme activities. They showed that the community had bacterial species capable of xylanase production. Portable metagenomic sequencing of genetically diverse RNA viruses on the Nanopore MinION direct from patient samples is shown to be a feasible methodology enabling a real-time field characterization of potential outbreaks (Kafetzopoulou et al., 2019). This technology had been employed to analyse RNAs of Lassa, hepatitis A and Ebola viruses (Kafetzopoulou et al., 2019). The same equipment can also be found at Plateau State University, Bokkos.

1.10 CHALLENGES OF MICROBIOME AND METAGENOMICS RESEARCH IN NIGERIA

1. Lack of collaboration

   Impactful research on the human microbiome and metagenomics research should be able to solve a problem and address issues of national interest. A review of research topics and project carried out in Nigeria had a touch of convenient basic research designed within infrastructural capacity in-country and the financial capability of the researcher to fulfil educational requirements or career promotion. Under such situations, the research outputs are felt by numbers of publication emanating from the research than translation into operational research or implementation. There is a paucity of data on collaborative research aimed to align activities to global national microbiome initiatives. Existing collaborating lack touch within. Moreover, research output does not reflect access to local resources and training.

2. Poor framework to support product development

   Biotechnology is the present-day science and related economic development. Molecular techniques are driving this industry. Knowledge of microbiomes and metagenomics has resulted in the discovery of novel natural products, new antibiotics, new molecules and bioactive molecules with new functions as well as new enzymes. It has defined genome/species and highlighted the interplay between humans and microbes and how microbial communities work. Bioinformatics makes use of computations, and a lot of software have been developed and in use. Human microbiome and metagenomics is herald by technology transfer, patenting, a massive spinoff of biotechnology companies, associated product and economic boost (Festel, 2013). In Nigerian system, a huge challenge exists on how to overcome the gap in technology transfer between innovation in basic academic, research and development (R&D) and on how research outcomes should be commercialized. There are no exiting records on any company whose products originate from microbial metabolites or genetic materials.

3. Inadequate funding

   Funding is a major drive in R&D. There is slow progress in the areas of microbiome and metagenomics research and consequently very little scientific contribution. Progress is hampered by low levels of available funding and investment. The cost of acquiring and maintaining expensive next-generation sequencers limits access to research into microbiome diversity in Nigeria. In other countries like the USA, China, the Netherlands and UK, funding through relevant stakeholders not only satisfies the needs for local research interest but is also expanded to other countries around the world. Affiliate countries are sponsored through grants to carry out metagenomics research addressing their research focus. There are centres in Nigeria involved in metagenomic research sponsored through this means. They have contributed in putting Nigeria into the limelight as publications emanating from such research carries names of Nigerian authors and affiliate centres (Dareng et al., 2016; Ayeni et al., 2018). This sort of collaboration keeps local capacity especially human resources busy by training and engaging researcher. Such is expected to continue at the close of projects following country ownership facilitated through the provision of adequate personnel and maintenance of infrastructure. This seems not to be so, as lack of funds, enabling environment, high-throughput technology and cyber infrastructures that support microbiome and metagenomics researches are not in place. Such limitation hampers design and implementation of large-scale genomic projects.

4. Training

   Capacity building assesses the level of technology, resources and ability to participate in genomics research. Considerable resources are needed for large-scale community genetic typing projects that explore environmental variation in human and environmental microbiome researches. To meet up with this, training is obtained through workshops, collaboration, exchange programmes, fellowship, doctorate, post-doctoral programmes and protocols from product manufactures. Training should be consistent with trends, guided by institutional policy and supported by funding. Lack of infrastructure and funding hampers training needed. The absence of metagenomic equipment limits the benefits of hands-on-training, thus hampering the acquisition of knowledge and skills needed.

5. Lack of infrastructure

   Metagenomics research involves high- and low-throughput machines and bioinformatics infrastructure. Equipment used are DNA Sequencer, Library Workstation, Ultrasonic Sample Processor, Fluorometer, Real-Time PCR and Pipetting Robot. The absence of these equipment increases the cost per sequenced base. It also aggravates limitation in human and computational resources to analyse metagenomics data. Data science is a new innovative approach that holds a large potential for creating business value (Kayser et al., 2018). Moreover, DNA codes for a lot of genetic information (Kozlov, 2018). These bioresources are guided, protected and controlled in other nations ethically and legally. Such wealth of big data that serve as raw material for R&D is not hosted physical in any hardware or electronically in the cloud. Online resources such as Metagenomics Rapid Annotation using Subsystem Technology (MG-RAST), Community Cyberinfrastructure for Advanced Microbial Ecology Research and Analysis (CAMERA) and Integrated Microbial Genomes and Microbiomes (IMG/M) are built to provide researchers with the resources to compare the metagenomic analysis of samples. Metagenomic data from a huge variety of environments are compared against host-associated metagenomic projects publicly available. Such cyber infrastructure is not yet available in Nigeria. Genomic information generated so far are uploaded on foreign databases. This has implication on ownership, patent, social security, health, protection of Nigerian bioresources and the environment. In addition to this, there are no standards and tools for present and future large-scale metagenome analyses that would turn the metagenomic data into knowledge.

6. Ethical issues

   Embracing human microbiome and metagenomic is challenged by ethical issues. Although there are guidelines available, the country is challenged by weak regulatory agencies to enforce regulated shipment of DNA materials out of Nigeria and fulfil the needs of the material transfer agreement recommended by the National Health Research Ethics committee.

1.11 PROSPECTS OF HUMAN MICROBIOME AND METAGENOMICS RESEARCH

a. **Computational biology:** Focus of microbiome and metagenomic research in Nigeria moved with the introduction of pyrosequencing and computational biology. Researchers are able to define compositional changes of microbiota and metabolic profiling (Waldor et al., 2015). So far, this research focus is able to define taxonomy and functional genes. This aspect should be inclusive of mechanisms mediating interactions between members of microbial communities and their hosts. In human microbiome, soil and marine metagenomics, huge input is required to catch up with current and future microbiome researches.

b. **Systems biology:** The era of systems biology-based approach is associated with the development of mathematical and computational models of specific components making up a community (Greenblum et al., 2012). This type of research provides a good understanding of the function and dynamics of microbiomes. There is a need to apply knowledge from microbiome research to the development of biomarkers that can be used to predict treatment responses towards diseases and disorders endemic to Nigeria and other African countries. The microbiome is linked with future health, making it a hot topic for health research. In the past two decades, the gut microbiome has been linked to several diseases and conditions, ranging from inflammatory bowel diseases to diabetes, autism, Parkinson's disease, anxiety and obesity to how individuals respond to certain drugs. Research into the vaginal microbiome is helping to define whether an HIV-prevention drug applied to the vagina would be effective (Waldor et al., 2015).

c. **Synthetic genomes:** Research into synthetic genomes resulted in the creation of novel microbes or synthetic microbiomes (Citorik et al., 2014). This offers a new way to bioengineering. Research into the processes on how to create new tools to manipulate the microbiota is still open for Nigerian researchers interested in microbiome and metagenomics research. Metagenomics study has shown a major impact on the way we study the microbiomes of environmental, clinical and engineered habitats. Technology has moved to high-throughput sequencing and new tools for analysing metagenomics. Biologically meaningful information is extracted from largely unassembled metagenomic datasets. Biological inferences can be made with increased confidence. More work is needed to examine the role of individual microbial populations in relation to other habitats in a community at genome, transcriptome and proteome levels. Future works are required to investigate more and generating multiple types of 'omic data, notably, metatranscriptomics, metaproteomics and metametabolomics, to characterize the microbiome.

d. **Bioeconomy:** Global human microbiome market is projected to reach up to USD 1,731 million in 2027 from USD 942 million in 2024, at a CAGR of 22.5% (Markets and Markets 2020). In Europe alone, metagenomics market is expected to reach US$ 89.7 million in 2027 from US$ 48.6 million in 2018 (https://www.researchandmarkets.com/r/8gjhml). Then at a CAGR growth rate of 7.4%, a greater percentage of this is expected to come from therapeutics, diagnostics, diseases and technology segment. Major activities to boost these areas are clinical trials for the development of target-specific, microbiome-based, bacteria-targeted therapies (Markets and Markets, 2020). As it is, Nigeria is not a market player as none of these products is manufactured locally. Rather, the country remains a consumer nation for its application and utilization of technology through importation. It is important therefore that Nigeria should invest massively to support R&D in universities and research institutes.

1.12 RECOMMENDATIONS

I. **Funding:** Microbiome and metagenomic research in Nigeria should be focus-driven to attract fundings that would support investments geared towards the promotion of research outputs. Benefits of microbiome and metagenomics in Nigeria are not in line with current trends in terms of collaboration which in most cases bring adequate funding to support research. There is a need to collaboration and align our activities to the global national microbiome initiatives.

II. **Research and Development:** Expand R&D in related government agencies to include microbiome and metagenomics studies and how to harness the educational and economical benefits. It would be important to track and analyse competitive developments such as acquisitions, partnerships and agreements on collaborations. These activities should be expanded to an aspect involved in monitoring product launches and expansions in the human microbiome market. At the national level, stakeholders should identify research interests/priority and work with relevant organizations to set out and achieve goals for national development. The benefits of investing in R&D out ways the challenges. Microbiome and metagenomics would contribute to solving problems. There is every hope that these fields would open the opportunity for the creation of jobs and diversification of the economy. This will also address issues of training need at undergraduate, postgraduate and post-doctoral levels.

III. **Need for a Bioresource Centre:** There is an urgent need to establish a bioresource centre with adequate infrastructure. The centre should be an integrated national centre for sample storage, sequencing, data storage and analysis as well as cloud computing for large-scale metagenome data analysis with guaranteed security. The centre should open up opportunities for collaborative research and coordination.

IV. **Collaboration:** Design of microbiome and metagenomic research encompasses varying skills and fields of expertise ranging from molecular biology to bioinformatics. Thus, it requires teams of experts and adequate funding to sustain relevant activities. It is important to collaborate with researches within Nigeria and promote access to local resources and training.

V. **Bioeconomy:** In order to harness the benefit of bioeconomy, there is an urgent need to support by special funding the spinoff of biotechnology companies as well as the acceleration of product and technology development. Such financing will help the companies to conduct research on microbiome-based products and metagenomic technology. Stakeholders should create awareness to enlighten academic researchers and open up the channel for industries to implement research-based cooperative agreements with academic spinoffs.

1.13 CONCLUSION

Researches on microbiomes and metagenomics have delivered many interesting basic findings and advanced technologies, enabling these disciplines to move from a marginalized position to becoming a beacon of hope with great potentials and many possibilities. The fields have spread out in many directions, with microbes being studied from all perspectives of life. The field is vast and virgin, and scientists/researchers in Nigeria are therefore encouraged to join the working group as we align our activities to the global national microbiome initiatives.

REFERENCES

Adelekan, O.A. and Esiobu, N. (2012). Bacterial succession studies during fermentation of African locust bean parkia biglobosa to iru using molecular methods. *British Biotechnology Journal* 2(1): 49–59.

Ajogbasile, F.V., Oguzie, J.U., Oluniyi, P.E., et al. (2020). Real-time metagenomic analysis of undiagnosed fever cases unveils a yellow fever outbreak in Edo State, Nigeria. *Scientific Report* 10: 3180.

Akubuenyi, F.C., Odokuma, L.O. and Nwaokorie, F.O. (2019). Metagenomics of bio slurry obtained from a biogas plant using next-generation DNA sequencing. *International Journal of Biotechnology Trends and Technology* 9(2): 12–17.

Amao, J.A., Barooah, M. and Omojasola, P.F. (2019). Comparative 16S rDNA metagenomics study of two samples of cassava peel heap from Nigeria and India. *Biotechnology* 9: 418.

Anika, O.C., Akin-Osanaiye, B.C. and Edet, U.O. (2020). Metagenomics in-sight into the microbial structural diversity of anaerobic digester utilising fruit waste as substrate. *Journal of Microbiology and Biotechnology and Food Sciences* 95: 965–969.

Anukam, K.C. and Agbakoba, N.R. (2017). A comparative study of the oral microbiome compositions of healthy postmenopausal, premenopausal and prepubertal Nigerian females, using 16SrRNA metagenomics methods. *Nigerian Journal of Clinical Practice* 20(10): 1250–1258.

Anukam, K.C., Agbakoba, N.R., Okoli, A.C. and Oguejiofor, C.B. (2019). Vaginal bacteriome of Nigerian women in health and disease: A study with 16S rRNA metagenomics. *Tropical Journal of Obstetrics and Gynaecology* 36: 96–104.

Awasthi, K.M., Ravindran, B., Sarsaiya, S., et al. 2020. Metagenomics for taxonomy profiling: Tools and approaches. *Bioengineered* 11(1): 356–374, doi: 10.1080/21655979.2020.1736238.

Ayeni, F.A., Biagi, E., Rampelli, S., Brigidi, P., Candela, M. and Turroni, S. (2018). Infant and adult gut microbiome and metabolome in rural bassa and urban settlers from Nigeria. *Cell Press* 23: 3056–3067.

Blaser, M., Cardon, J., Cho, Z.G., et al. (2016). Toward a predictive understanding of Earth's microbiomes to address 21st century challenges. *mBio* 73: e00714–16.

Bukar, M., Sodipo, O., Dawkins, K., Ramirez, R., Kaldapa, J., Tarfa, M. and Esiobu, N. (2019). Microbiomes of top and sub-layers of semi-arid soils in North-Eastern Nigeria are rich in firmicutes and proteobacteria with surprisingly high diversity of rare species. *Advances in Microbiology* 9: 102118, doi: 10.4236/aim.2019.91008.

Chikere, C.B., Mordi, I.J., Chikere, B.O., Selvarajan, R., Ashafa, T.O. and Obieze, C.C. (2019). Comparative metagenomics and functional profiling of crudeoil-polluted soils in Bodo West Community, Ogoni, with other sites of varying pollution history. *Annals of Microbiology* 69: 495–513.

Chikere, C.B., Surridge, K.J., Cloete, T.E. and Okpokwasili, G.C. (2011). Phylogenetic diversity of dominant bacterial communities during bioremediation of crude oil- polluted soil. *Revista Ambiente and Água* 6: 61–76.

Citorik, R.J., Mimee, M. and Lu, T.K. (2014). Bacteriophage-based synthetic biology for the study of infectious diseases. *Current Opinion in Microbiology* 19: 59–69.

Dareng, E.O., Ma, B., Famooto, et al. (2016). Prevalent high-risk HPV infection and vaginal microbiota in Nigerian women. *Epidemiology and Infection* 144: 123–137.

Datta, S., Rajnish, K.N., Samuel, M.S., et al. (2020). Metagenomic applications in microbial diversity, bioremediation, pollution monitoring, enzyme and drug discovery. A review. *Environmental Chemistry Letters* 18: 1229–1241.

Dudas, G., Carvalho, L.M. and Rambaut, A. (2017). Virus genomes reveal factors that spread and sustained the Ebola epidemic. *Nature* 544: 309–3152017.

Edet, U.O. and Antai, S.P. (2018a). Correlation and distribution of Xenobiotics genes and metabolic activities with level of total petroleum hydrocarbon in soil, sediment, and estuary water in the Niger Delta. *Asian Journal of Biotechnology and Genetic Engineering* 11: 1–10.

Edet, U.O., Antai, S.P., Brooks, A.A. and Asitok, A.D. (2017a). An overview of cultural, molecular and metagenomics techniques in description of microbial diversity. *Journal of Advances in Microbiology* 72: 1–19.

Edet, U.O., Antai, S.P., Brooks, A.A. and Asitok, A.D. (2017b). Metagenomic assessment of antibiotics resistance genes from four ecosystems in the Niger Delta area of Nigeria. *Asian Journal of Biotechnology and Genetic Engineering* 11: 1–10.

Edet, U.O., Antai, S.P., Brooks, A.A. and Asitok, A.D. (2018b). Evaluation of microbial diversity of epipellic and benthic sediment using cultural and metagenomics techniques. *Asian Journal of Environment and Ecology* 6(4): 1–13.

Etebu, E., Torunana, J.M.A. and Parker, M. (2018). Metagenomic analysis of bacterial community associated with postharvest Irvingia species fruit wastes. *Microbiology Research International* 62: 7–15.

Ezekoye, C.C., Chikere, C.B. and Okpokwasil, G.C. (2018). Field metagenomics of bacterial community involved in bioremediation of crude oil-polluted soil. *Journal of Bioremediation and Biodegradation* 95: 449.

Fasesan, D., Dawkins, K., Ramirez, R., Rasheed-Jada, H., Onilude, A., Nash, O. and Esiobu, N. (2020). Analysis of a tropical warm spring microbiota using 16S rRNA metabarcoding. *Advances in Microbiology* 10: 145–165.

Festel, G. (2013). Technology transfer models between industrial biotechnology companies and academic spin-offs. *Industrial Biotechnology* 9(5): 252–257.

Gevers, D., Knight, R., Petrosino, J.F., Huang, K., McGuire, A.L., Birren, B.W., Nelson, K.E., White, O., Methé, B.A. and Huttenhower, C. (2012). The Human Microbiome Project: A community resource for the healthy human microbiome. *PLoS Biology* 10(8): e1001377.

Gilbert, J.A. and Lynch, S.V. (2019). Community ecology as a framework for human microbiome research. *Nature Medicine* 256: 884–889.

Greenblum, S., Turnbaugh, P.J. and Borenstein, E. (2012). Metagenomic systems biology of the human gut microbiome reveals topological shifts associated with obesity and inflammatory bowel disease. *Proceedings of the National Academy of Sciences* 109: 594–599.

Groer, M.W., Luciano, A.A., Dishaw, L.J., Ashmeade, T.L., Miller, E. and Gilbert, J.A. (2014). Development of the preterm infant gut microbiome: A research priority. *Microbiome* 2: 38, doi: 10.1186/2049-2618-2-38.

Handelsman, J., Rondon, M.R., Brady, S.F., Clardy, J. and Goodman, R.M. (1998). Molecular biological access to the chemistry of unknown soil microbes: A new frontier for natural products. *Chemistry and Biology* 510: R245–R249.

Jain, M., Olsen, H.E., Paten, B. and Akeson, M. (2016). The Oxford nanopore MinION: Delivery of nanopore sequencing to the genomics community. *Genome Biology* 171: 239.

Kafetzopoulou, L.E., Pullan, S.T., Lemey, P., et al. (2019). Metagenomic sequencing at the epicenter of the Nigeria 2018 Lassa fever outbreak. *Science* 363: 74–77.

Kanehisa, M. and Goto, S. (2000). KEGG: Kyoto encyclopedia of genes and genomes. *Nucleic Acids Research* 28(1): 27–30.

Kayser, V., Nehrke, B. and Zubovic, D. (2018). Data science as an innovation challenge: From big data to value proposition vict. *Victoria, Technology Innovation Management Review* 8(3): 16–25.

Kozlov, N.N. (2018). The study of the secrets of the genetic code. *Journal of Computer and Communication* 6: 64–83.

Kuczynski, J., Stombaugh, J., Walters, W.A., Gonzalez, A., Caporaso, J.G. and Knight, R. (2012). Using QIIME to analyze 16S rRNA gene sequences from microbial communities. *Current Protocols in Bioinformatics.* Chapter 10, Unit10.7–10.7.

López-García, P. and Moreira, D. (2008). Tracking microbial biodiversity through molecular and genomic ecology. *Research in Microbiology* 1591: 67–73.

Ma, J., Prince, A.L., Bader, D., et al. (2014). High-fat maternal diet during pregnancy persistently alters the offspring microbiome in a primate model. *Nature Communication* 5: 3889.

Markets and Markets (2020). Human Microbiome Market – Global Forecast to 2027 | By Product, Application, Disease & Research Technology | Markets and Markets. www.marketsandmarkets.com/pdfdownloadNew.asp?id=37621904/, Accessed 13/6/2020.

Mboto, C.I., Edet, U.O., Mbim, E.N., et al. (2018). Human microbiome diversity: Implication in health, disease and applications. *World News of Natural Sciences* 21: 98–117.

Mercer, B., Dawkins, K., Meday, L. and Esiobu, N. (2020). Metagenomic shotgun sequencing provides prevalence data for pathogens, and source-tracking indices useful in public health risk assessment of environmental waters. *Journal of Geoscience and Environment Protection* 8: 115–129.

Ndiokwere, C., Olise, N.A., Nmewurum, V., Omoregie R., Agbakoba R.N., Anukam and K.C. (2019). 16S rRNA metagenomics of seminal fluids from medical microbiology laboratory in a tertiary hospital, Southern Nigeria. *Journal of Medical Laboratory Science* 29(3): 86–109.

Nwaokorie, F.O., Maiga, M., Patel, P., Kanki, P. and Ogunsola, F. (2020). Distinct difference between microbiome of rural and urban, population in Lagos State, Nigeria. *African Journal of Medicine and Medical Science* 49: 363–378.

Ogbulie, T.E. and Nwaokorie, F.O. (2016). Molecular diversity of microbes with probable degradative genes in agricultural soil contaminated with bonny light crude oil. *Journal of Ecosystem and Ecography* S5: 002, doi: 10.4172/2157-7625.S5-002.

Olukunle, O.F. and Boboy, B. (2013). The molecular succession of bacterial community of crude oil polluted soil and water samples from the Niger Delta, Nigeria. *British Journal of Applied Science and Technology* 3: 777.

Oulas, A., Pavloudi, C., Polymenakou, P., et al. (2015). Metagenomics: Tools and insights for analyzing next-generation sequencing data derived from biodiversity studies. *Bioinformatics and Biology Insights* 9: BBI-S12462.

Salam, L.B., Ilori, M.O., Amund, O.O., LiiMien, Y. and Nojiri, H. (2018). Characterization of bacterial community structure in a hydrocarbon-contaminated tropical African soil. *Environmental Technology* 397: 939–951.

Salam, L.B., Obayori, O.S., Nwaokorie, F.O., Suleiman, A. and Mustapha, R. (2017). Metagenomic insights into effects of spent engine oil perturbation on the microbial community composition and function in a tropical agricultural soil. *Environmental Science and Pollution Research* 24: 7139–7159.

Shrinivas, N.S., Padmaja, P.S. and Krishnaraj, P.U. (2019). Soil Metagenomics: Concepts and Applications, Metagenomics - Basics, Methods and Applications, Wael N. Hozzein, Intech Open, doi: 10.5772/intechopen.

Thompson, L.R., Sanders, J.G., McDonald, D., et al. (2017). The Earth Microbiome Project Consortium. A communal catalogue reveals Earth's multiscale microbial diversity. *Nature* 551: 457–463.

Udofia, U.U., Edet, U.O. and Antai, S.P. (2018). Potential benefits of applying "Omics" Technology in cleaning up crude oil spillages in the Niger Delta Region of Nigeria. *Advances in Research* 15(2): 1–8.

Udotong, I.R., Uko, M.P., and Udotong, J.I.R. (2015). Microbial diversity of a remote aviation fuel contaminated sediment of a lentic ecosystem in Ibeno, Nigeria. *Environmental & Analytical Toxicology* 5: 6, doi: 10.4172/2161-0525.1000320.

Udotong, I., Uko, M. and Udotong, J. (2017). Use of 16S rDNA sequencing to determine procaryotic diversity of a remote aviation fuel-polluted lentic ecosystem in Ibeno, Nigeria. *Journal of Environmental & Analytical Toxicology* 7: 493.

Uko, M.P., Udotong, I.R. and Umana, S.I. (2019). Homology-dependent metagenomics study of agricultural soils in Akwa Ibom State, Nigeria. *Journal of Biology and Innovation* 81: 54–65.

Waldor, M.K., Tyson, G., Borenstein, E., et al. (2015). Where next for microbiome research? *PLoS Biology* 13(1): e1002050.

Zehr, J.P., Weitz, J.S., and Joint, I. (2017). How microbes survive in the open ocean. *Science* 357 (6352): 646–647.

2 Microbial Biotechnology
Expanding the Frontiers of Microbiomes for Innovation in Agriculture and Medicine

Olawole O. Obembe
Covenant University

Nwadiuto (Diuto) Esiobu
Applied Biotech Int'l Nigeria Ltd
Florida Atlantic University

O. S. Aworunse
Covenant University

Nneka R. Agbakoba
Nnamdi Azikiwe University

CONTENTS

2.1 Introduction 18
2.2 The Plant Microbiome 20
2.3 Ecology and Structure of Plant Microbiome 20
  2.3.1 Below-Ground (Rhizosphere) Microbiome 21
  2.3.2 Above-Ground (Phyllosphere) Microbiome 21
  2.3.3 Plant Internal Tissue (Endosphere) Microbiome 22
2.4 Complementary Functional Capacities of the Plant Microbiome 22
2.5 The Deployment of the Collaborative Functional Capabilities of Plant Microbiomes Could Spinoff Next-Generation Agro-Innovations 22
  2.5.1 Microbial Inoculants 25
  2.5.2 In Situ Microbiome Engineering 26
2.6 Artificial Ecosystem Selection of Microbiome 26
2.7 Microbial-Based Approach 26
2.8 Biochemical Approach 27
2.9 Molecular Approach 27
2.10 Expanding the Frontiers of Plant Microbiomes 28
2.11 Human Microbiome 28
  2.11.1 Composition 28

DOI: 10.1201/9781003180241-2

2.11.2 Early Development of the Human Microbiome ... 29
2.11.3 Microbiome and Its Correlations with Human Health ... 30
2.12 Microorganisms from Selected Body Sites and Their Association with Health ... 30
2.12.1 Oral Microbiome ... 30
2.12.2 Gut Microbiome ... 31
2.12.3 Vaginal Microbiome ... 32
2.12.4 Skin Microbiome ... 33
2.13 Role of the Human Microbiome in Some Metabolic Conditions ... 34
2.14 Metabolic Disorders ... 34
2.14.1 Obesity ... 34
2.14.2 Type 2 Diabetes ... 34
2.14.3 Neurologic Disorders ... 35
2.14.4 Depression ... 35
2.14.5 Amyotrophic Lateral Sclerosis (ALS) ... 37
2.14.6 Parkinson's Disease (PD) ... 37
2.15 Autoimmune Disorders ... 38
2.15.1 Type 1 Diabetes Mellitus (T1DM) ... 39
2.15.2 The Systemic Lupus Erythematosus (SLE) ... 39
2.15.3 Rheumatoid Arthritis (RA) ... 40
2.15.4 Inflammatory Bowel Disease (IBD) ... 40
2.16 Microbiome and the Role of Probiotics ... 41
2.16.1 Vaginal Infection ... 41
2.16.2 Metabolic Syndrome ... 42
2.17 Innovative Applications of Human Microbiome in Medicine ... 42
2.18 Skin Microbiome Applications ... 43
2.18.1 Forensic Applications ... 43
2.18.2 Therapeutic Applications ... 43
2.18.3 Cosmetics and Anti-aging ... 44
2.18.4 Oral Microbiomes Applications ... 45
2.18.5 Gut Microbiome Applications ... 46
2.19 Conclusions ... 47
References ... 47

2.1 INTRODUCTION

It is now evident that plants and animals are a composite of not only the products of their genomes but also of the genomes of associated symbiotic microorganisms that reside in and on them (Bulgarelli et al., 2013). Due to this polygenomic nature, several authors have considered plants and animals as holobionts or meta-/supraorganisms (Bosch & Mcfall-Ngai, 2011; Rajpoot et al., 2018) that depend partly on their cognate microorganisms for specific characteristics and functions (Mendes et al., 2013). Following the successful completion of the Human Genome Project, there also arose the need to query microbial assemblages associated with specific organs in different higher organisms (Davies, 2001; Compant et al., 2019).

The last decade has witnessed a surge in microbiome research. Consequently, it has become apparent that microbiota performs crucial functions in determining human health (Rajpoot et al., 2018), conferring resilience to environmental stress in plants, enhancing plant growth, as well as protecting plants against pathogens and pests (Sessitsch et al., 2019). Whereas the microbiota is defined as the set of microorganisms in a given habitat, the microbiome is the collection of microbes and their genomes in specific habitat (Bulgarelli et al., 2013). The term holobiont refers to the association between a host and all its microbiota, which both interact and perform synergistically (Sessitsch et al., 2019). Efforts to decipher the constituents of microbial communities and the specific metabolic interrelationships of the same with their associated host have broadened microbial biotechnology research from traditionally investigating individual genomes to studying microbial communities and their interactions in different host ecosystems (Turnbaugh et al., 2007). This has birthed the field of metagenomics, which allows in situ identification of all the genomes of the many individual organisms in these communities, as well as their gene expression and metabolic products (Turnbaugh et al., 2007; Turner et al., 2013). Rapid advances made in systems biology-based approaches such as high-throughput sequencing technologies, artificial intelligence, data mining, advanced cloud computing, and bioinformatics have substantially sharpened our current understanding about the composition, ecology, functional capabilities, and dynamics of microbial assemblies in their associated hosts (Turner et al., 2013; Bai et al., 2019).

Plant microbes are critical influencers of soil function (soil biology) and crop productivity (Berendsen et al., 2012), and have gained significant recognition lately (Lebeis et al., 2012). Despite improved chemical solutions and agronomy, the agriculture industry is beset with mounting pressures such as an ever-increasing population that is anticipated to reach an estimated 9 billion by the year 2050 (Raines, 2018), loss of cultivable land owing to urbanization and soil degradation, decreasing water availability, and 30% yield loss to pest damage and disease (Qiu et al., 2019; Sessitsch et al., 2019). Furthermore, experts have speculated climate change to adversely affect crop productivity (Lobell & Field, 2007). Mobilizing microbiome functions to improve crop yields holds tremendous potential to mitigate the current challenges of land degradation, reduced crop yield, and food security (Qiu et al., 2019). Specifically, with a global burden to boost food production while lowering applications of agrochemicals and fertilizers, an appropriate knowledge on mobilizing the plant microbiome, for instance, for nitrogen fixation, becomes increasingly imperative (Hirsch, 2018).

The diverse community of microbes hosted in the human body collectively forms the human microbiota (Clemente et al., 2012). These microbial symbionts, mostly resident in the large intestine of the gut, by far surpass the number of human germ and somatic cells by a factor of ten (Matamoros et al., 2013). The human microbiota is estimated to encode more than 100 times the number of genes in the human genome (Magalhaes et al., 2007). It confers characteristics that humans do not need to evolve on their own (Turnbaugh et al., 2007). In collaboration with their host, the microbiome interactively performs many biological processes (Sommer & Bäckhed, 2013). These include metabolic fermentation and absorption of undigested polysaccharides, harvest, and storage of energy (Gill et al., 2006), as well as immune system signaling to enhance maturation of immune cells and healthy development of immune function (Chow et al., 2010). Accumulating evidence shows that the microbiota has profound implications for human health and disease (Rajpoot et al., 2018). Disruption in the community structure of the microbiota, caused by eco-physiological changes of a niche can induce altered interactions with the host, leading to the development of many diseases (Sommer & Bäckhed, 2013). Insights into the microbiota, communications between all its components and tissues in the human host, will be critical for inventing predictive model diagnosis and developing microbiota-based interventions for the treatment of diseases linked with microbiota imbalance (Clemente et al., 2012).

This chapter discusses microbiota and microbiome compositions in human and plant holobionts, their beneficial interactions with the associated hosts, as well as research frontiers currently shaping our knowledge about the establishment and sustenance of the host-microbiota relationships. The chapter further highlights how an understanding of these interactions could be crucial to spawning the next wave of innovations in healthcare and agriculture.

2.2 THE PLANT MICROBIOME

Nearly all plant tissues harbor an assortment of microbial populations ranging from archaea, bacteria, protists, viruses to fungi (Parasuraman et al., 2019). These microbial communities, which can exist, survive, and cooperatively function with various plant tissues such as shoots, roots, leaves, seeds, and flowers, are referred to as the plant microbiome or "phytobiome" (Parasuraman et al., 2019). This complex network of microbial populations performs crucial roles in host's nutrient assimilation, resilience to abiotic and biotic stresses, and metabolic capabilities (Mohanram & Kumar, 2019). Considering that these microbes are regularly recruited by their host to perform critical functions, plant-associated microbiota is regarded as the accessory plant genome (Sessitsch et al., 2019). Plant microbiome structure is determined by diverse elements like plant developmental stage, age, health, and cultivar or species (Berg et al., 2015b). Specific microbial communities may colonize plants stochastically. Nevertheless, an increasing body of evidence suggests that plant core microbiome (species-specific microbial communities) are established consistently and reproducibly in a spatiotemporal manner, and partly across organs, not only at phylum level but also at higher taxonomic resolution (Vorholt et al., 2017). The profusion of each phylum, classes, or genera varies to some extent due to differences in plant genotypes and the environment (Vorholt et al., 2017). Soil is the primary source of microbes, which shapes the plant-associated microbiome (Pineda et al., 2017).

2.3 ECOLOGY AND STRUCTURE OF PLANT MICROBIOME

Different compartments of the same plant interactively play host to distinct microbiomes, which colonize three major plant regions: (1) below ground (rhizosphere), (2) above ground (phyllosphere), and (3) internal tissue (endosphere) (Parasuraman et al., 2019). Microbiota in these environments can initiate symbiotic, commensal, or pathogenic associations of varying intimacies with their host plants (Dawkins & Esiobu, 2017, 2018). Whereas microbes that reside in the phyllosphere and rhizosphere (those that adhere to plant tissue) are regarded as epiphytes, those that live in the endosphere (within plants) are considered endophytes (Turner et al., 2013).

2.3.1 Below-Ground (Rhizosphere) Microbiome

The soil milieu surrounding the plant root surface is regarded as the rhizosphere (Qiu et al., 2019). Considered as a hotbed for microbial activities compared to bulk soil, it is one of the most cryptic ecological niches (Compant et al., 2019). Plants impact the rhizosphere via deposition of exudates from root tip cells (primarily antimicrobial compounds, fatty acids, plant growth hormones, phenolics, vitamins, organic acids, and amino acids), sloughed cells, and mucilage (Rout, 2014; Compant et al., 2019). Discharges from the root are usually species-specific (Rout, 2014) and are critical influencers of rhizospheric microbiome structure (Turner et al., 2013). These exudates facilitate communication between plants and rhizosphere microbiota to assist in disease resistance and physiological performance (Korenblum et al., 2020). By utilizing plants' exudate as a source of nutrients, microbes benefit from the association (Rout, 2014). The root microbiome comprises rhizosphere-inhabiting bacteria (and their genes) that can proliferate within the soil and initiate collaborative functions with their plant host (Hao & Xiao, 2017). Mostly derived by horizontal transfer from the surrounding soil vicinity (Compant et al., 2019), the rhizosphere microbiota incorporates fungi, oomycetes, bacteria, protozoans, nematodes, viruses, archaea, and algae (Mendes et al., 2013). Bacteria are the most prevalent with the preponderance of *Proteobacteria* (α and β classes), *Actinobacteria*, *Planctomycetes*, *Bacteroidetes*, *Verrucomicrobia*, and *Firmicutes* (Compant et al., 2019). Different cultivars of the same plant species maintain a high level of selectivity for rhizosphere microbiota (Korenblum et al., 2020). Rhizospheric microbial diversity fluctuates depending on soil environment, plant genotype, and stage of development (Korenblum et al., 2020). The majority of the rhizospheric microbes are growth-promoting and do not pose any harm to their plant hosts (Hao & Xiao, 2017).

2.3.2 Above-Ground (Phyllosphere) Microbiome

The phyllosphere encompasses different aerial plant surfaces such as flower surfaces (asthenosphere), stem surface (caulosphere), leaf surface (phylloplane), fruit surface (carposphere), and apoplast of leaf tissue (Parasuraman et al., 2019). While phyllosphere microbiota represents a broad spectrum of microorganisms (Stone et al., 2018) such as filamentous fungi, yeast, bacteria, algae, and less commonly, nematodes and protozoa (Lindow & Brandl, 2003), it is generally dominated by bacteria. *Proteobacteria* (α and γ classes) are the most frequently encountered phylum, with *Firmicutes*, *Bacteroidetes*, and *Actinobacteria* are also well represented (Bodenhausen et al., 2013). The phyllosphere microbiota typically derives from the air, soil, and seed and becomes attuned for association with plant tissues where host genotype, environmental, edaphic, and cultivational practices determine community structure (Compant et al., 2019). Generally, phyllosphere bacterial communities are less species-rich relative to the rhizosphere or soil (Stone et al., 2018). Compared to the rhizosphere, the phyllosphere is nutrient-poor (Turner et al., 2013) and is affected by fluctuations of temperature, moisture, and radiation (Bringel & Couée, 2015). The distribution and establishment of microbes on the leaves are not uniform, and are somewhat influenced by morphological features like stomata, trichomes, and veins (Lindow & Brandl, 2003).

2.3.3 Plant Internal Tissue (Endosphere) Microbiome

Microbes whose life cycles are hosted at least in part within plant tissues are known as endosphere microbiomes (Monteiro et al., 2013). Whereas endophytic microbiomes are typically non-pathogenic, eliciting no visually observable symptoms, they include dormant microbes that could become pathogenic depending on host genotype or environmental factors (Monteiro et al., 2013). *Rhizobia* and *Firmicutes*; *Actinobacteria* and α, β, and γ *Proteobacteria*; *Firmicutes*; and *Actinobacteria* are ubiquitous endophytes. Endophytes are thought to be a subset of the rhizospheric microbial communities (Compant et al., 2010) and mainly hosted in dying or dead cells, as well as intercellular apoplast (Bulgarelli et al., 2012). Albeit many endophytes are systematically translocated through the xylem to specific plant regions like fruits, stem, and leaves, they can also gain entry into plant tissues via aerially located flowers and fruits. In nature, endophytes can passively access plant endosphere through injuries on roots or root tips, site of emergence of lateral roots, as well as by vegetative propagation means (pieces of the stem). Within plants, endophyte coloniality and transmissibility are shaped by conditions such as plant resource allocation (Compant et al., 2019).

2.4 COMPLEMENTARY FUNCTIONAL CAPACITIES OF THE PLANT MICROBIOME

The recognition that microbial assemblages are associated with almost every host and play vital roles in well-being and productivity throughout the entire life cycle of the hosts has been of remarkable progress in biological sciences. Many functions of the plant microbiome are critical for the host (Berg et al., 2015b). These are considered in Table 2.1.

2.5 THE DEPLOYMENT OF THE COLLABORATIVE FUNCTIONAL CAPABILITIES OF PLANT MICROBIOMES COULD SPINOFF NEXT-GENERATION AGRO-INNOVATIONS

The enhancement of crop yield via modification of plant microbiome is an antiquated practice that predates 300 BC (Versey, 2003). A knowledge of plant-microbiome collaborations is pivotal to developing strategies for reconfiguring rhizospheric, endospheric, and phyllospheric microbiota to provide beneficial plant services. Since the release of the first commercial bioinoculant, Nitragin (a culture of *Rhizobium*), which was patented in 1986 (Finkel et al., 2017), research efforts have been geared toward the development of microbial products for agriculturally important legumes (Mohanram & Kumar, 2019). The advent of the "next-generation" and even "third-generation" technologies (Rout, 2014) is already providing platforms to decipher the complementary roles performed by plant microbes. This could be instrumental in spawning the next wave of agro-innovations via the deployment of microorganisms beneficial to crop productivity and growth. In the subsequent sections, we highlight research frontiers in this regard and their capacity for improved crop yield.

TABLE 2.1
Functional Interactions of Plant Microbiomes

Microbiome Niche in Plant Host	Complementary Services Provided	References
Rhizosphere	PGPR belonging to the *Actinomycetes*, *Firmicutes*, *Protobacteria*, and *Actinobacteria* confer suppression against plant pathogens by disrupting virulence factors or by releasing a broad spectrum of molecules with antagonistic properties (antifugal, antibacterial, antiviral, insecticidal, and nematicidal activities).	Loper et al. (2012), Turner et al. (2013), Rout (2014), Berg et al. (2015b), Mauchline and Malone (2017), Mohanram and Kumar (2019)
	PGPR such as symbiotic root-nodule associated *Rhizobium* spp. and free-living Azotobacter are involved in nitrogen fixation in soil. Many rhizosphere-inhabiting prokaryotes solubilize phosphorus-containing compounds, thereby enhancing their bioavailability to plants.	
	Rhizospheric microbiomes promote plant growth and stress tolerance by manipulating phytohormones, especially gibberellins, auxins, and ethylene.	
	Rhizospheric *Pseudomonas* spp. secrete a varied assortment of natural compounds, many of which function to destroy or restrict plant pathogens. These include the antifungal molecule diacetylphloroglucinol (DAPG), siderophores, phenazines, pyoluteorin and pyrrolnitrin, hydrogen cyanide. It eventuates that diacetylphloroglucinol-producing *Pseudomonas* spp. alter transcriptional patterns in the PGPR, *Azospirillum brasilense*, thus enhancing the expression of genes that influence root proliferation and plant growth.	
	Many rhizospheric prokaryotes produce N-acyl-L-homoserine lactones (AHLs)-degrading enzymes known to inhibit quorum sensing in pathogens.	
	Plant growth promoting bacteria (PGPB) confers immunity against a wide range of phytopathogens by eliciting induced systemic resistance (ISR) via the release of volatile organic compounds (VOCs).	
	Rhizosphere microbiome also perform pivotal roles in the decomposition of organic matter decomposition, thereby promoting soil fertility and ultimately improving plant yield.	

(Continued)

TABLE 2.1 (*Continued*)
Functional Interactions of Plant Microbiomes

Microbiome Niche in Plant Host	Complementary Services Provided	References
Phyllosphere	Phyllospheric microbiome produces fluence-sensing rhodopsins and proton pumps which assist the plant in energy capture and metabolism. Microbiota hosted in the phyllosphere accentuates plant resistance to pathogen invasion either via regulation of host defense or other sorts of interactions with pathogens. Several foliar microbiotas can fix carbon and nitrogen by assimilating plant-derived amino acids and ammonium, this being the primary process of supplying nitrogen to plants growing in humid tropical ecosystems. Other valuable functions of the microbiota hosted on plant surfaces include the production of plant hormones.	Turner et al. (2013), Rudrappa et al. (2008), De Vrieze et al. (2018), Compant et al. (2019)
Endosphere	By expressing genes and proteins that drive nitrogen fixation, endophyte-inhabiting diazotrophs help to fix and transfer nitrogen to their plant host in substantial amounts. Endophytic bacteria enhance plant defense against phytopathogenic invasion. Plant growth-promoting endophytes (*Rhizobia*, *Bacillus* spp.) synthesize and release phytohormones such as gibberellins, auxins (indole-3-acetic acid (IAA)), and cytokinins that stimulate plant growth by altering endogenous hormone level in collaboration with a plant host. Additionally, PGPB belonging to *Bacillus* spp., *Athrobacter* spp., and *Athrobacter* spp. produce an enzyme, 1-aminocyclopropane 1-carboxylic acid (ACC) deaminase, which enhances host tolerance to environmental stress by interfering with the synthesis of the stress hormone, ethylene. The role of endophytes belonging to *Enterobacter* and *Serratia* in the inhibition of take-all disease (*Gaeumannomyces graminis*) that causes significant losses in cereals have also been documented.	Turner et al. (2013), Mauchline and Malone (2017), Durán et al. (2018), Singh et al. (2019), Compant et al. (2019)

2.5.1 Microbial Inoculants

Bioformulations incorporating single strain or consortia (Bellabarba et al., 2019) of varying plant growth-promoting organisms capable of fixing, mobilizing, solubilizing, or decomposing nutrient sources when applied via soil or seed, to promote nutrient availability and assimilation by plants, are regarded as microbial inoculants (Mohanram & Kumar, 2019). Also included in this category are products that can be classified as plant strengtheners, biofertilizers, phytostimulators, and biocontrol agents, or biopesticides (Lugtenberg & Kamilova, 2009). Many plant growth-promoting microbes have been applied to the soil, on seeds, and other plant propagules as bioinoculants to improve plant performance and stimulate indigenous beneficial rhizospheric bacteria (Bhattacharyya & Jha, 2012). An example is the exploitation of plant growth-promoting rhizobacteria (PGPR) for enhanced crop yield, and *Bacillus thuringiensis* and *Trichoderma* spp. as biocontrol of plant pests and pathogens (Qiu et al., 2019). Over the last three decades, many Rhizobial bio-inoculants for lacing legume seeds have been patented and are now available in the market (Ab Rahman et al., 2018; Bellabarba et al., 2019). Introduction of genetically engineered microbial strains as designer probiotics (Mendes et al., 2013) tailored to specific soil environments (Schlaeppi & Bulgarelli, 2015) can help plants overcome pathogens by pervading the rhizosphere and attaining cell densities above a particular limit required for pathogen infection at a specific time and place (Mendes et al., 2013). More so, microbial inoculants have been documented to promote plants' innate resistance to diseases, thus showing suitability as substitutes for pesticides, fungicides, and antibiotics. Likewise, biofertilizers incorporating live microbes that can be introduced to plants, soil, or seeds are now widely deployed in organic farming to improve the plant's innate ability to acquire nutrients, thereby reducing the costs associated with repeated fertilizer use. It is touted that developing nations in Africa and Asia can profit from multi-strain biofertilizers formulated from rhizosphere soil with up to 10% anticipated increase in crop productivity (Qiu et al., 2019). Specific molecular signatures of the microbiome of healthy plants can be exploited to formulate microbial inoculants capable of producing desirable outcomes when applied to seeds. Recognizable examples include strains of *Streptomyces* that distinctly modulate plant growth and yield in response to drought and bacterial endophytes that promote maize seedling fitness under various stress states (Riley, 2018).

Additionally, the use of microbial consortia comprising assemblages of diverse microorganisms that can function synergistically or complementarily is receiving considerable attention (Mendes et al., 2013). The consortia are formulated to include an aggregate of fungal or bacterial species to cover a wide range of applications and soil environments (Bellabarba et al., 2019). Owing to their ability to work synergistically in eliminating toxins and exchanging nutrients, microbes within a consortium are better adapted to withstand biotic and abiotic stresses (Sekar et al., 2016). The application of *Rhizobia* in collaboration with other PGPR may promote bacterial performance. For example, a combination of *Rhizobium* with *Bacillus* strains have been demonstrated to improve root morphology and exacerbate nodule development in some legumes (Bellabarba et al., 2019). More so, microbial consortia could be mobilized to defend against fungal pathogens (Hacquard, 2018). For example, microbial consortia from the families *Pseudomonadaceae* and *Comamonadaceae* have been reported to inhibit the occurrence of disease in maize and sugar beet eventuated by the fungal pathogens, *Fusarium* and *Rhizoctonia* (Trivedi et al., 2017).

2.5.2 In Situ Microbiome Engineering

It is now known that soil microbes can over time adapt to their host crop, thereby fostering an enhanced plant-microbe interactivity (Berendsen et al., 2012). Several studies have substantiated the overarching roles played by native microbes in pathogenesis in plants (Bulgarelli et al., 2013). Hence, exploiting the innate competence of the autochthonous plant microbiome can enable the selection of novel and superior microbial functions (Qiu et al., 2019). In situ microbiome engineering is a novel research frontier in agriculture (Qiu et al., 2019), which alters plant's naturally occurring microbiome to improve function (Mueller & Sachs, 2015). This strategy produces soil/plant-optimized microbes harboring microbiomes that can be applied as inoculum for diverse crops in varied soils (Ab Rahman et al., 2018). Plant native microbiota can be manipulated on-site by mimicking naturally occurring ecological mechanisms via the following approaches: (1) artificial ecosystem selection of microbiome (AESM), (2) microbial-based, (3) biochemical, and (4) molecular-based (Qiu et al., 2019).

2.6 ARTIFICIAL ECOSYSTEM SELECTION OF MICROBIOME

Root-associated microbiota is representative of the plant microbiome and the environment. Functionally, the root microbiome antagonizes plant pathogens and impacts plant physiology (Agler et al., 2016). Consequently, engineering root-microbiota via microbiome transfer can be employed to manipulate microbiome structure and interactions for enhanced plant health, growth, and overall fitness. AESM is an emerging area of synthetic ecology, which involves the transplantation of pathogen-suppressive soils (and their core microbiome) to pathogen-burdened soils to mitigate plant diseases (Gopal et al., 2013; Turner et al., 2013). This method is in many ways similar to human fecal microbiota transplantation (FMT) in which beneficial microbiota from a healthy donor is transferred to patients with gastro-intestinal disease to achieve re-establishment of healthy gut microbiomes (Foo et al., 2017). AESM relies on the principle of cross-compatibility of the core microbiomes of plant species that are phylogenetically unrelated (Gopal & Gupta, 2016). This strategy has been deployed to suppress plant diseases, including tobacco black root rot, potato common scab, and sugar beet infection (Gopal et al., 2013). More so, microbiome transfer from sympatric soil has shown the potentials to improve biomass under drought conditions and promote drought tolerance in *Arabidopsis thaliana* (Zolla et al., 2013).

2.7 MICROBIAL-BASED APPROACH

This approach is based on the ecological phenomenon of priority effects. It targets hub microbiota (components of plant microbial community that can establish strong collaborations and orchestrate microbial community assembly, via the release of signals) for manipulation on the field. The microbial-based strategy is achieved by timing and managing hub microbiota establishment, growth, and proliferation to promote the arrival of other functional members of the microbial community during the early phases of plant growth. Following specific steps, the hub microbiota is identified, isolated, and applied as parental floral sprays to impregnate seeds with specific hub microbiota-harboring microbiomes with superior capacity to recruit functional microbes before seedling establishment. Using a microbial-based strategy, Xue et al. (2015) developed the biofertilizer, BIO, to reconfigure banana rhizospheric microbiome structure and subsequently dampened the incidence of Panama disease. For further details on the microbial-based strategy of in situ microbiome engineering, refer to Qiu et al. (2019).

2.8 BIOCHEMICAL APPROACH

This technique harnesses in situ molecular signals released by both microbes and plants to recruit and perpetuate hub microbiota or other functional microorganisms (Qiu et al., 2019). For the fact that plants can deploy beneficial rhizospheric microbiota via root exudates, manipulating the composition of the same, both quantitatively and qualitatively, is a core strategy to reconfigure rhizosphere microbiota. "Rhizosphere engineering" (Bakker et al., 2012) is a novel biochemical strategy that relies on modification of root exudation patterns in the rhizosphere to establish a bias for beneficial microbes. This approach demands extensive knowledge of the chemical variants of rhizodeposits as well as the molecular communications in the rhizosphere (Mendes et al., 2013). Rhizosphere engineering reshapes microbial populations by attenuating the growth of specific organisms (Mohanram & Kumar, 2019). Besides root exudates, volatile organic compounds released by plants can attract microbiota that confers suppressiveness on soils from distal locations, thus showing the capacity for engineering soil microbes to promote plant defense (Schulz-Bohm et al., 2018). Biochemical strategies allow candidate root metabolites that stimulate the proliferation of specific rhizospheric microbial components to be identified, isolated, synthesized, and applied to accentuate the plants' competence to recruit and maintain the activities of beneficial rhizospheric microbes (Qiu et al., 2019). It is now known that the application of root exudates as prebiotic can trigger the establishment of modular microbiomes that protect plants against pathogens and stress (Gundel et al., 2011). Foliar application of plant-derived compounds modifies microbial structure in favor of plant growth-promoting and antifungal bacteria (belonging to the genus *Bacillus*, *Acinetobacter*, *Pseudomonas*, and *Pantoea*). This approach has been applied to enhance microbiome biodiversity in lettuce (Luziatelli et al., 2019).

2.9 MOLECULAR APPROACH

There are indications that modern plant breeding may have selected against plant characteristics that are critical for harboring and supporting beneficial microbiota (Smith et al., 1999). By exploiting the association between the host plant genome and microbiome, it is possible to breed plants that harbor superior microbiome using quantitative trait loci (QTL) mapping. QTL mapping can be leveraged to identify genes or loci that exert control over the interactivities between beneficial microbiomes and crops. By deploying genetic engineering or traditional breeding approaches, these QTLs can be harnessed subsequently to breed improved crop varieties, which have the capacity for recruiting beneficial microbial communities (Qiu et al., 2019). One of the pioneering studies to determine the molecular basis of plant communication with pathogen-inhibiting rhizobacteria was carried out by Smith et al. (1999). Using QTL, the authors identified three loci that correlate with disease suppression by a strain of *Bacillus cereus* in inbred tomato lines. Their results indicated that genetic variation in host plant species could be leveraged to promote beneficial associations between plants and rhizospheric microbiomes (Mendes et al., 2013). In furtherance, genetically manipulated potatoes have been documented to release molecular signals that interrupt quorum sensing in pathogenic bacteria (Bauer & Mathesius, 2004) through the expression of a lactonase gene. This trait conferred a higher level of resilience against the pathogen *Pectobacterium carotovorum* (Dong et al., 2001). Moreover, GM potato plants with enhanced pectate lyase (Wegener, 2001) and 5-O-glucosyltransferase (Lorenc-Kukula et al., 2005) production inhibited *Pectobacterium carotovorum.*

2.10 EXPANDING THE FRONTIERS OF PLANT MICROBIOMES

A burgeoning population, coupled with an increasing number of plant pathogens, puts global food security under threat (Ab Rahman et al., 2018). Breeding plants that can select for and maintain beneficial microbiomes via appropriate exudate secretion (microbiome-optimized plants) (Trivedi et al., 2017), engineering microbiomes for incorporation as microbial consortia with superior adaptability to support plants in varied soil types (plant-optimized microbiome), and more efficient use of single or multiple bacterial species as microbial biopesticides and biofertilizers (personalized plant medicine), are evolving as favorable options to enhance productivity and complement plants in the fight against pathogens (Ab Rahman et al., 2018). Translation and integration of the approaches, as earlier mentioned (in this section) for practical applications, may open up new vistas to "personalized" agriculture aimed at enhancing and sustaining global food production in the future to come. The development of personalized models is already yielding results for Big Pharma. The agricultural biotechnology industry could also adopt these new customized models to produce premium quality crops (Stokes & McCourt, 2014).

2.11 HUMAN MICROBIOME

2.11.1 Composition

The microbiome inhabits the human body; on the skin, in the oral cavity, in the respiratory tract, in the gut, the genital tract, and even in sites like the blood, previously considered as sterile environments (Proal et al., 2014). The human body has several habitats or niches, with each habitat playing host to a unique microbiome composition. The diversity of microbiomes varies from habitat to habitat, with the immune system of individuals restricting the microbiota to their unique niches in the body. The composition of the human microbiome differs from person to person, with several factors playing interrelated roles. Environmental changes, for example, can lead to prior adaptation that was favorable to an individual becoming adverse. It can be as a result of leaving a local setting to an urban setting or even traveling out of a warm climatic environment to a much colder climate. Such factors can lead to alteration in microbial composition. Diet can also play a role, likewise factors like antibiotics consumption, cosmetics types—soap, cream, lotion, toothpaste, bathing water, drinking water. Genetics is also another factor as the genetic make-up of humans differs from one person to another.

The composition of microbiomes of healthy humans is not exactly known; because what is fit for one individual may not be for the other. However, some microbiota are known to be predominantly present in some habitats where they most often play protective roles of identifying and destroying invading microbes that come into their habitats. According to Dominguez-Bello et al. (2019), bacteria like *Staphylococcus*, *Streptococcus*, *Actinomyces*, *Veillonella*, *Fusobacterium*, *Porphyromonas*, or *Treponema* species are dominant in the oral cavity. Bacterial phyla like Firmicutes and Bacteroidetes and genera like *Bacteroides*, *Prevotella*, *Ruminococcus*, *Bifidobacterium*, *Streptococcus*, *Enterobacteria*, *Enterococcus*, *Lactobacillus*, and *Akkermansia* are present in the gastro-intestinal tract. *Lactobacillus* species abound in the women's genital tract, while on the skin are found bacteria like *Firmicutes* (*Staphylococcaceae*), *Bacteroidetes*, *Proteobacteria*, *Actinobacteria* (*Propionibacteriaceae* and *Corynebacteriaceae*) and the fungus *Malassezia* species. Other habitats like the nasal and ocular also have their peculiar microbiome. Apart from bacteria that occupy more than 90% of the total human microbiome, fungi, viruses, and archaea are also present in humans, although in much reduced numbers.

Compositions of human microbiomes are much more than the above-listed organisms, as scientists have come to understand in the past decade. Currently, due to the use of next-generation sequencing techniques like the 16S ribosomal RNA metagenomics, the entire microbiome of particular microbial communities can be harvested from their natural environments, and their DNA sequences analyzed without needing to cultivate the microorganisms. As this is capable of revealing the natural culturable bacteria, the unculturable bacteria, the anaerobes, and even some yet-to-be-named detected bacteria. Thus, in a single test, universal coverage of all bacteria in a community can be attained.

2.11.2 Early Development of the Human Microbiome

A recent discovery that placentas are free of bacteria suggests that babies pick microbiome during and after birth (de Goffau et al., 2019). However, the birth mode is a significant factor that affects the microbiota type of an infant at the early stage of life. During vaginal birth, the new born becomes exposed to the mother's fecal and vaginal microbiota and gets colonized with typical vaginal bacteria (e.g., *Prevotella*, *Atopobium* spp.). It takes about 24 hours for babies born through a cesarean section to become populated with bacteria genera such as *Propionibacterium* and *Staphylococcus* generally found on the mother's skin (Dominguez-Bello et al., 2019). The composition of the fetal microbiome in preterm babies also differs according to their gestational weight gain (Antony et al., 2015).

Feeding type also contributes to the early development of the human microbiome. The distinction in the microbial composition of formula- or breast-fed infants is quite clear. A higher proportion of *Bifidobacteria* and *Lactobacillus* were observed in breast-fed children, while formula-fed infants have a higher prevalence of *Clostridiates* and *Proteobacteria* (Bokulich et al., 2016). The diversity and bacterial richness of formula-fed babies decreased after the first year of life. Another factor affecting the infant microbiome is prenatal maternal exposure to antibiotics, which alters the diversity of both maternal and neonatal microbiota (Stiemsma & Michels, 2018).

The microbiome is rapidly populated within the first 3–4 weeks and through the first year of life. After weaning, the gut microbiota becomes firmly established, leading to a lifelong microbiome signature in healthy individuals. However, factors like diet, antibiotics intake, as well as other environmental factors continue to change the microbiome, and this may continue throughout life. Hence as humans develop into childhood and adulthood, so does their microbiomes diversify in richness and are more stable except when dysbiosis occurs as a result of diseases.

2.11.3 Microbiome and Its Correlations with Human Health

In humans, microorganisms impart health, infectious diseases as well as some metabolic conditions. The human microbiome project has offered the scientist the opportunity to study the entire microbiota of different body sites to know the precise microbial composition of the community. With this, all the microorganisms found in both healthy and diseased humans are not only identifiable but also distinguishable with their specific roles established.

In the human body, microbial communities abound, and they play critical roles, some of which include: food digestion, generation of energy, detoxifying chemicals in the environment, barrier functions as the first line of defense on the skin and mucous membrane, preventing the growth of harmful microorganisms and disease progression, maintaining the host immune system, moderating cytokines through the pro- and anti-inflammatory processes, and promoting the colonization of beneficial microorganisms (Kilian et al., 2016).

2.12 MICROORGANISMS FROM SELECTED BODY SITES AND THEIR ASSOCIATION WITH HEALTH

2.12.1 Oral Microbiome

The human oral cavity, which is a significant gateway to the human body, is made up of different niches like the throat, cheek, tongue, teeth, saliva, and floor of the mouth—all of which allow for microbial colonization (Benn et al., 2018). External factors such as oral hygiene practices, types of food intake, oral fluids intake, temperature and humidity, and also exposure to chemicals in the environment affect the oral micro-habitat (Stanaway et al., 2016). Escapa et al. (2018) reported that a good number (57%) of the bacteria in the oral cavity had been named, about 13% cultivated while the remaining 30% remained uncultured.

The microbiome composition of the oral cavity is studied using metagenomic tools. In a study, the oral bacteriome of randomly selected healthy individuals showed that there was no significant difference between males and females on the relative abundance of the genera and species colonizing the oral cavity of the tested population. Generally, in both males and females, *Firmicutes* (43.6%) were the most abundant phyla followed by *Proteobacteria* (33.62%), *Bacteroidetes* (9.67%), *Actinobacteria* (8.48%), *Fusobacteria* (4.31%), Spirochaetes (0.18%), *Candidatus*, *Saccharibacteria* (0.04%), *Tenericutes* (0.03%), *Synergistetes* (0.02%), and *Acidobacteria* (0.01%). Others that recorded less than 0.01% were *Streptophyta*, *Chloroflexi*, *Chlamydia*, and *Cyanobacteria*. The abundant genera detected were *Streptococcus*, *Haemophilus*, and *Neisseria*, while at the species taxonomic level, the most abundant core species identified were *Haemophilus parainfluenzae* and *Haemophilus influenza* (Anukam et al., 2018).

Though most of these phyla contain a healthy bacteriome, some contain genus and species like *Porphyromonas gingivalis*, *Treponema denticola*, and *Tannerella forsythia*, which have been described as disease-causing agents and implicated in periodonititis. Other organisms, linked with periodontitis, are classes *Clostridia*, *Erysipelotrichia*, and *Negativicutes* (Griffen et al., 2012); the genera *Fusobacterium*, *Prevotella*, and *Synergistes*; as well as archaeal species *Methanobrevi bacteroralis*, *Methanobacterium curvum/congolense*, and *Methanosarcina mazeii* (Matarazzo et al., 2011). Fungal genera have also been reported from the cavity genera like *Aspergillus*, *Candida*, *Penicillium*, *Schizophyllum*, *Gibberella*, and *Rhodotorula* (Peters et al., 2017).

2.12.2 Gut Microbiome

The gut microbiome comprises the genetic make-up of the entire microbes that live in the digestive tract of humans. Microbes in the human gut make up to 100 trillion cells, ten times the number of human cells, and they include bacteria, fungi, viruses, and protozoa, although most often, the focus is only on the bacterial composition (Landman & Quevrain, 2016). The majority of the gut microbiota belong to four main phyla, namely, *Actinobacteria*, *Bacteroidetes*, *Firmicutes*, and *Proteobacteria*. A mutualistic relationship exists between gut microbiota and humans to maintain homeostasis (Yap & Marino, 2018).

A metagenomic technique using the 16S ribosomal RNA gene sequencing to analyze the gut microbiota from stool samples have revealed a wide variety of bacteria, especially the unculturable ones. Healthy individuals have phyla like *Bacteroidetes*, *Firmicutes*, *Proteobacteria*, *Caldithrix*, *Cyanobacteria*, *Planctomycetes*, *Spirochaetes*, *Tenericutes*, *Chloroflex*, *Chrysiogenetes*, *Chlorobi*, *Fusobacteria*, *Thermotogae*, *Euryarchaeota*, *Deferribacteres*, *Verrucomicrobia*, *Thermi*, and *Synergistic* (Belizário et al., 2018). From these, some essential genera like *Ruminococcus*, *Prevotella*, and *Bacteroides* were also quite prominent. Some of the prevalent species found in standard stool samples include: *Bacteroides dorei*, *Bacteroides fragilis*, *Alistipes finegoldii*, and *Alistipes onderdonkii* as well as some species from the genera *Coprococcus*, *Eubacterium*, and *Ruminococcus*. Another prominent bacterium is *Faecalibacterium prausnitzii*, a Gram-negative obligate anaerobe, which is taxonomically in the *Clostridia* order of phylum *Firmicutes* seem to be present in the gut of healthy humans; it may act as a biomarker of a healthy gut (Oshim, 2020).

Some intestinal bacteria, as etiologic agents, cause some gastro-intestinal infections. Bacteria like *Helicobacter pylori* is involved in stomach ulcer disease and chronic gastritis, as well as being a risk factor in gastric cancers. Bacterial genera like *Streptococcus*, *Prevotella*, *Neisseria*, *Haemophilus*, and *Porphyromonas* also cause the development of gastritis in humans (Althani et al., 2016). Bacteria from the intestinal microbiome somewhat cause extra-intestinal disorders. An imbalance of the gut microbiota might underlie a broad spectrum of human diseases, such as immune, metabolic as well as neurological disorders.

2.12.3 Vaginal Microbiome

The healthy vaginal microbiota is predominantly Gram-positive rods, with a prevalence of *Lactobacillus* species, which represent 80%–95% of the resident bacteria at this site (Cribby et al., 2008). These bacteria possess some properties that not only enable them to colonize the mucous membranes but also prevent the establishment or excessive proliferation of potentially pathogenic microorganisms (Livengood, 2009). The pathogenic bacteria found in the lower female genital tract include *Streptococcus*, *Propionibacterium*, *Eubacterium*, *Bacteroides*, *Atopobium*, *Prevotella*, *Mycoplasma*, *Ureaplasma*, and *Gardnerella vaginalis*, among others (Martin, 2008). When the lactobacilli concentration in the vagina reduces, these bacteria may overgrow to the extent that they dominate the environment as opportunistic pathogens. Other pathogenic organisms like fungi, protozoa, and viruses could be causative agents, but the focus here is on the bacteriome of the vaginal tract.

Most bacterial vaginal infections are a result of dysbiosis in the original microbiota whereby the predominant, protective *Lactobacilli* species gave way to the vaginal commensals, which become opportunistic pathogens. This event is typical in the case of bacterial vaginosis (BV), a polymicrobial vaginal infection. Currently, using the 16S rRNA metagenomics technique, the microbiome of the vagina has been extensively studied to reveal the presence of many more unculturable bacteria and anaerobes.

Anukam et al. (2019) conducted a study on Nigerian women with and without BV using the 16S rRNA metagenomics technique. They reported that some of the BV-causing bacterial species detected included *Gardnerella vaginalis*, *Corynebacterium* species, *Prevotellabivia*, *Mycoplasma hominis*, *Sneathia sanguinegens*, *Prevotella amnii*, *Atopobium vaginae*, *Ureaplasma urealyticum*, *Megasphaera* sp. UPII 199-6, and *Collinsella aerofaciens*. The three most abundant bacteria in samples predicted to be BV positive were *Gardnerella vaginalis*, *Sneathia sanguinegens*, and *Atopobium vaginae*. The microbial diversity score of one of the control subjects indicated that the subject's vaginal microbiome was more diverse in terms of species richness than that of the BV subjects. This could be a result of the presence of the high occurrence of *Lactobacillus crispatus*, which is beneficial in the vagina. *Lactobacilli* species were found dominating the vagina of women without

BV, thus indicating that healthy Nigerian women had *Lactobacillus* genera as the predominant microbiota.

2.12.4 Skin Microbiome

Next to the human intestine, the human skin is the second largest organ of the body. As an outer covering, it serves as a protective physical barrier and one of the first-line defenses of the body to the external environment, protecting the body against harmful pathogens. Skin is one of the human body sites that possess a healthy microbiota. A variety of microorganisms colonize the skin, some of which may be harmless or beneficial to the host. The skin has different niches and different adapted microorganisms depending on the skin thickness, skin folds, the moisty state of the niche, and the density of hair follicles present. Microorganisms on the skin differ from person to person and from niche to niche.

Consequent to the use of 16S ribosomal RNA metagenomic sequencing, many unculturable bacteria are detected and added to the vast array of the skin microbiome. Most skin bacteria exist in four different phyla, namely, *Actinobacteria*, *Firmicutes*, *Bacteroidetes*, and *Proteobacteria*. These four dominant phyla also correspond to the microbiota found on the oral cavity as well as the gastro-intestinal tract, although with varying proportions. Members of the phylum *Actinobacteria* are prevalent on the skin, while those of the phyla *Bacteroidetes* and *Firmicutes* are prevalent in the gastro-intestinal tract.

A study on the microbiome of the axilla skin samples of young adults in Nigeria revealed the presence of 27 phyla with the four prominent ones as follows: *Actinobacteria* (60%), *Firmicutes* (32%), Proteobacteria (5%), and Bacteroidetes (2%). The genera *Corynebacterium* (56%), *Staphylococcus* (21%), and *Anaerococcus* (6%) were predominantly the common genera detected in the participants, along with their numerous species. The genera *Anaerostipes* and *Anaerovibrio* were mostly exclusive to the males, while *Olsenella*, *Vitreoscilla*, *Rheinheimera*, *Sorangium*, *Ferrimicrobium*, *Solibacillus*, and *Rummeliibacillus* were the predominant exclusive genera to the females. *Corynebacterium* and *Staphylococcus* species were among the most abundant bacteria colonizing moist areas of the skin, such as the axilla, which suggests that these organisms thrive better in very humid areas (Mmewurum, 2020).

The skin microbiome and diseases revealed that the microbiome of normal healthy skin has a high diversity and high interpersonal variation as against that of a disease condition. Microbiota compositions of diseased lesioned skin in cases of atopic dermatitis and psoriasis showed distinct differences compared with healthy skin. Non-bacterial microorganisms like fungi have also caused skin infections, e.g., *Malassezia* spp.

Skin-associated diseases, linked to specific and robust microbiome dysbiosis, include acne vulgaris caused by *Propionibacterium acnes* and atopic dermatitis (AD), a type of eczema with *Staphylococcus aureus* been mostly implicated (Dessinioti & Katsambas, 2010). Skin commensals like *Staphylococcus aureus*, *Enterococcus* spp., or *Pseudomonas aeruginosa* commonly infect burn wounds.

2.13 ROLE OF THE HUMAN MICROBIOME IN SOME METABOLIC CONDITIONS

Recent findings showed that changes to gut microbiota play an essential role in the pathogenesis of some disease conditions which may be metabolic, neurologic, and autoimmune disorders.

2.14 METABOLIC DISORDERS

2.14.1 OBESITY

Obesity is a nutritional disorder, which is characterized by excessive body mass index (BMI). It is a threat to health as it may adversely affect the individual's health. In obesity, factors like dysbiosis in the gut microbiota, wrong diet, and environmental factors, alter health condition. Previously known risk factors for obesity include diet, lifestyle, and socioeconomic status. Currently, however, the gut microbiota has emerged as a relatively new factor with a much more special role to play. The involvement of the gut microbiome in the intestinal integrity, energy balancing, and providing immunity against hostile microbial invaders gives them control of the host's overall health status. The permanent role played by gut microbes on human health has been linked to the balanced gut microbiota. As such, any imbalance, or dysbiosis in the gut microbiota, has been linked to multiple conditions, one of which is obesity (Tseng & Wu, 2019).

The microbiomes of lean and obese people vary in the sense that obese people possess lower microbial diversity as against the richer diversity of lean people. Metagenomic analysis on 16S rRNA of the human gut microbiome has shown that the microbiome of obese persons has an abundance of phylum *Firmicutes* and a reduced abundance of the *Bacteroidetes*. So, an alteration in the gut environment that warrants changes in the *Bacteroidetes/Firmicutes* ratio provides genetic material that enables individuals to have the capacity to harvest energy from the diet. This disturbance consequently promotes lipogenesis, leading to an increase in the quantity of the lipid droplets and eventually to excess weight gain.

Aside *Firmicutes* and *Bacteroidetes*, many other bacteria play roles in gut dysbiosis that will eventually lead to various metabolic disorders. Among them are families *Prevotellaceae* and *Enterobacteriaceae*, which predominate obesity subjects, and *Christensenellaceae* found more in subjects with low BMI (Goodrich et al., 2014). Also implicated in obesity are genera like *Lactobacillus*, *Bacteroides*, and *Roseburia*, whose abundance positively correlated with BMI (Patil et al., 2012).

2.14.2 TYPE 2 DIABETES

Diabetes mellitus is one of the metabolic disorders of significant public health importance. It presents with elevated blood glucose due to the body's inability to produce enough insulin to meet its needs. Current research is gearing toward the search for the role of the gut microbiome in the development of diabetes. The presence of many unculturable bacteria is ubiquitous in the gut microbiome of humans, as revealed by the 16S metagenomic sequencing; as such, they play causal roles in diabetes.

The dysbiosis associated with type 2 diabetes mellitus (T2DM) presents as very low in both the diversity and richness of the microbiota. *Actinobacteria*, *Bacteroidetes*, *Firmicutes*, *Proteobacteria*, and *Verrucomicrobia* are among the most abundant phyla in fecal samples. The relative abundances of *Bacteroidetes* and *Firmicutes* were higher in non-DM than in T2DM subjects. Larsen et al. (2010) reported that changes in the composition of intestinal microbiota were associated with T2DM compared to non-diabetic controls. These authors showed that the *Bacteroides*: *Firmicutes* ratio was raised in T2DM, and this correlated with raised plasma glucose on Oral Glucose Tolerance Test.

On the contrary, in a study conducted on Chinese subjects, the microbiome diversity did not reveal any difference between the T2DM and the non-DM subjects (Qin et al., 2012). However, differences exist in the bacterial composition and function of the gut microbiota whereby increased opportunistic pathogens, butyrate-producing bacteria, and species with potential for mucin-degradation and sulfate-reduction were observed. These authors further reported that certain genes belonging to the species *Akkermansia muciniphila* were enriched in T2DM, whereas genes belonging to *Haemophilus parainfluenzae* were enriched in control subjects.

Also, comparing T2DM subjects to non-diabetic subjects, showed that diabetic patients had a decrease in butyrate-producing bacteria such as *Roseburia intestinalis*, *Eubacterium halii*, and *Faecalibacterium prausnitzii*, and increases in *Escherichia coli*, *Lactobacillus gasseri*, *Streptococcus mutans*, and some *Clostridium* species. Oshim et al. (2020) reported that the high abundance of *Prevotella copri* and low abundance of *Akkermansia muciniphila* in T2DM represents intestinal pathobionts signatures associated with T2DM development. These findings go to show that there is obvious relationship between gut microbiota and T2DM.

2.14.3 Neurologic Disorders

Studies show that the human microbiome ecosystem can interfere with the brain's development, and so, disruption in the human microbiome may affect the course of mental disorders. Since the gut contains roughly the same number of neurons as the spinal cord, it has been considered as the second brain of the human being. Therefore, phrases like "gut feeling," "gut thinking," "gut instinct," and "trusting the gut" are quite apt in everyday English language. Studies have shown that the gut microbiome in humans has an impact on the functions of the central nervous system (CNS) through identified pathways called the microbiome-gut-brain (MGB) axis. Here, we discuss a few neurologic disorders below.

2.14.4 Depression

Depression is a common mental disorder that contributes significantly to the overall global burden of disease. Patients are always in a depressed mood and lack interest in daily activities entirely. Factors such as gut dysbiosis and emotional stress can contribute to modulating the risk of depression. An inflammatory reaction of the body against the lipopolysaccharide of Gram-negative bacteria and bacterial translocation was proposed as critical factors that participate in the development of clinical depression (Berk et al., 2013). So, apart from the significant role of low-grade inflammation in the pathophysiology of depression, disturbance of microbiota and its association with its host have become another new pathway suspected of inducing an inflammatory response in depression (Maes et al., 2013). Changes in the microbiota of an individual are capable of affecting both the immune and nervous systems of that individual. This can reduce the person's ability to cope with physical and psychological stress and so becomes venerable to stressful everyday activities. Therefore, with secure communication at the MGB axis, an association between the gastro-intestinal system and the nervous system is established.

The development of the human MGB axis begins early in life. When one is going through stress, the brain reactions can change gut gastro-intestinal functions, including gut microbiome composition. Early life disruption of the MGB axis has been associated with later life neuropsychiatric disorders such as depression (Rogers et al., 2016). This, therefore, suggests a link between the gut microbiome, neurodevelopment, and depression.

Studies have revealed a characteristic gut microbiota in depression to include both high level of *Bacteroidales* and low *Lachnospiraceae* composition, and both an increased *Alistipes* population and *Enterobacteriaceae* level and low grade of *Faecalibacterium* in depression (Naseribafrouei et al., 2014). It has been shown that clinical depression is accompanied by elevated levels of immunoglobulin A or M, or both against Gram-negative bacteria belonging to human gut microbiota (Wiest, 2015), such as *Hafnia alvei*, *Pseudomonas aeruginosa*, *Morganella morganii*, *Proteus mirabilis*, *Pseudomonas putida*, *Citrobacter koseri*, or *Klebsiella pneumoniae* (Maes et al., 2013).

A Belgian study that compared microbes from depression patients and other participants observed that two kinds of microbes, *Coprococcus* and *Dialister*, were conspicuously missing from the microbiomes of the depressed subjects, but not from those of healthy controls subjects (Valles-Colomer et al., 2019). This finding was also confirmed when data from participants from Dutch were examined, and the same group of bacteria was also found missing (Pennisi, 2019). Additionally, the Gut-brain module analysis of fecal metagenomes identified the microbial synthesis potential of the dopamine metabolite 3,4-dihydroxyphenylacetic acid as correlating positively with mental quality of life. It indicated a potential role of microbial γ-aminobutyric acid production in depression (Valles-Colomer et al., 2019).

In an experiment involving the use of mice, Zheng et al. (2016) were able to show that depression-like behaviors can be transferred from humans to germ-free mice via fecal microbial transplantation. The transplanted feces, obtained from depressive patients, were able to induce depression-like expression in the germ-free mice, thus establishing an animal model of humanized depression. This fact confirms that an association exists between depression and the gut microbiome. Given the fact that gut microbiome plays roles in depression, the use of probiotics, like the fecal transplants, will help restore or alter the gut microbiome in depressed people.

2.14.5 Amyotrophic Lateral Sclerosis (ALS)

ALS, also called Lou Gehrig's disease, is a neuromuscular disease characterized by progressive death of motor neurons and muscle atrophy. The condition is accompanied by de-regulated energy metabolism as well as systemic and CNS-specific inflammation. These are potential pathogenic factors found to interact with gut microbiota (Brenner et al., 2017). ALS is a neurodegenerative disease with a pathophysiology that may be related to the gastro-intestinal tract. Gut microbiota, therefore, may have an impact on brain activity through the gut-microbiota-brain axis under both physiological and pathological conditions.

Zhou and Sun (2015) discovered that a link between intestinal homeostasis and ALS progression is through a leaky gut, which contributes to a change in the microbiome environment that leads to the reduction of beneficial bacterial products or enhancement of toxic bacterial products. Either of these events can lead to further worsening of the gut leakage, thereby promoting local and systemic inflammatory response, which can have a significant impact on the neuromuscular system and the entire body's physiological homeostasis in ALS patients.

Different studies on ALS have revealed varying results in the microbiome of subjects. The relative abundance of beneficial microorganisms decreased, while those of harmful microorganisms showed an increased tendency in patients with ALS. Compared to healthy subjects, the relative abundance of gut bacteria revealed a higher rate of *Firmicutes* as against the lower rate obtained for *Bacteroidetes* in patients with ALS, thus indicating that the gastro-intestinal tract health of patients with ALS was significantly influenced by this disease (Fang et al., 2016). It was reported that the genera obtained in ALS and non-ALS individuals were *Anaerostipes*, *Akkermansia*, *Bacteroides*, *Blautia*, *Bifidobacterium*, *Escherichia*, *Faecalibacterium*, *Fusicatenibacter*, *Megamonas*, *Shigella* and *Streptococcus* (Zhai et al., 2019). Though the beneficial bacterium, *Faecalibacterium*, was found in both healthy and ALS patients, the prevalence obtained for the ALS patients was lower (4.8%) than the 9.9% obtained from the healthy controls (Miquel et al., 2013). The relative abundance of harmful bacteria like genus *Dorea* was enhanced in ALS patients.

In contrast, that of beneficial bacteria such as genera *Anaerostipes*, *Lachnospiraceae*, and *Oscillibacter* showed a declined trend in the gastro-intestinal tract of patients with ALS (Fang et al., 2016). A recent study has discovered nicotinamide (NAM), a metabolite of a bacterium, *Akkermansia muciniphila*, as the causal link between the bacterium and ALS (Blacher et al., 2019). They reported that when given alone, NAM entered the bloodstream of the mice from the gut and alter gene expression in the brain.

2.14.6 Parkinson's Disease (PD)

PD or simply Parkinson's is a long-term neurogenerative disorder of the CNS that affects movements. Symptoms of PD include difficulty in walking, tremors, presence of alpha-synuclein (αSyn) as well as inflammatory cytokines within the brain (JNPD,

2017). The production of short-chain fatty acids (SCFAs), like acetate and butyrate, by gut bacteria, when they breakdown dietary fiber, can also activate an immune response in the brain.

Working with a mouse model, researchers used mice that overproduce αSyn and display symptoms of Parkinsonism on the one hand; on the other hand, they used germ-free mice that lack entirely gut bacteria. To measure the motor skills of the experimental animals, both groups of mice were made to perform several tasks that included crossing a beam, running on a treadmill, and descending from a pole. The germ-free mice were found to have performed better than those with a complete gut microbiome (JNPD, 2017). In another experiment, these researchers used fecal samples from both PD patients and healthy controls and transplanted into two sets of germ-free mice. The mice that received the transplant from PD patients exhibited symptoms of PD as well as the increased levels of SCFAs found in their feces. In contrast, the mice that received the fecal transplant from healthy controls did not exhibit any PD symptoms.

Many studies have been carried out on the microbiome of PD patients and healthy controls to reveal varying results. From the fecal samples of PD patients, Keshavarzian et al. (2015) reported a decrease in some butyrate-producers like *Faecalibacterium*, *Blantia*, and *Roseburia* as well as in the genus *Dorea*. Another study said that the infection of *Clostridium difficile* in the gut caused dysbiosis that resulted in the increased abundance of the family *Lactobacilliceae* and the decrease of the genera *Dorea*, *Roseburia*, *Blautia*, and *Faecalibacteria* (Antharam et al., 2013). Genus *Akkermansia* is prevalent in PD patients than in healthy controls. The family *Bifidobacteriaceae* showed varying results in PD patients. Where some studies reported a high abundance of the family in the mucosa and feces of PD patients compared to controls, others reported low abundance (Li et al., 2017). The families *Enterobacteriaceae* and *Lactobacilliaceae* vary in abundances, with some researchers reporting more and others reporting less in the PD patients they studied. The gut microbiota thus plays essential roles in the diagnosis of PD. New evidence suggests that PD starts in the gut and spread to the brain by the propagation of pathologic α-synuclein protein via the vagus nerve (Kim et al., 2019).

2.15 AUTOIMMUNE DISORDERS

Autoimmune disorders are those diseases whereby an individual's immune system attacks self-tissues. In an autoimmune disease, the immune system does not recognize self-cells, which it sees as foreign, and so releases autoantibodies that attack the healthy cells. Apart from aberrant autoantibodies production, genetic, immunologic, and environmental factors also play contributory roles in autoimmune diseases. A significant function of the gut microbiota is the maintenance of homeostasis of the human immune system, and so, any dysbiosis in the gut can adversely affect the host. Dysbiosis of the gut microbiome, which can come in the form of excessive growth of potentially pathogenic organisms, loss of beneficial organisms, or the entire loss of microbial diversity, can induce autoimmune disease in people (De Gruttola et al., 2016). Several autoimmune diseases are known, and some of them include type 1 diabetes mellitus (T1DM), rheumatoid arthritis (RA), systemic lupus erythematosus (SLE), and inflammatory bowel disease.

2.15.1 Type 1 Diabetes Mellitus (T1DM)

T1DM, which is also known as juvenile diabetes, is a chronic disease condition found mostly in children and young adults. Its main features are the production of little or no insulin by the pancreas and increased blood glucose levels. Changes in the composition of gut microbiota are implicated in T1DM etiology (Zhou et al., 2018). The possible pathogenic mechanisms by which gut dysbiosis can occur in T1DM subjects are through gut leakiness and immunological deregulation.

The phyla *Actinobacteria*, *Bacteroidetes*, and *Proteobacteria* were more abundant in Type 1 diabetes subjects while *Fusobacteria*, *Tenericutes*, *Verrucomicrobia*, and *Firmicutes* were more abundant in the control group (Brown et al., 2011). Giongo et al. (2011) reported that the autoimmune microbiome for T1DM might be quite different from that found in healthy children. While the microbiome of healthy children is richer, more diverse, and more stable, that of autoimmune children are less stable and less diverse. Reports also have it that while bacterial diversity increased in control subjects, it remained constant in T1DM cases, thereby indicating that an immunological response has taken place in autoimmune children, which might have reduced the microbial diversity in them. Reports showed that as children become autoimmune, there was a decline in the phylum *Firmicutes* as against the increase in phylum *Bacteroidetes*. The opposite was noticed among healthy children as *Bacteroidetes* decreased as against the *Firmicutes* that increased in abundance. Giongo et al. (2011) noted that in the phylum *Bacteroidetes*, the most prevalent species was *Bacteroides ovatus*, which contributed about a quarter of the entire constituents of the phylum in the T1DM subjects.

Bacteria genera like *Akkermansia* and *Prevotella*, as well as *Clostridium* clusters IV and XIVa, were reduced in T1DM children but were more prevalent in healthy children. Other investigators also reported a decrease in the genus *Bifidobacterium* in T1DM subjects (Zhou et al., 2018). All these findings go to show that the gut microbiota has a definite role to play in T1DM.

2.15.2 The Systemic Lupus Erythematosus (SLE)

The SLE, also simply called "Lupus," is a heterogeneous autoimmune disease with a wide range of clinical and serological manifestations. It can affect the brain, joints, skin, blood cells, kidneys, heart, and lungs. The disease may be mild or severe and marked by relapses and remissions. The condition affects more women than men often. The pathogenesis of SLE is poorly understood and may involve hormonal factors, genetic and environmental factors like infection, and drugs (Zhang et al., 2014). Like in other autoimmune diseases, the gut microbiota has been suspected of playing a contributory role as etiologic agents in the pathogenesis of SLE. Reports of many studies show that dysbiosis of the intestinal microbiota is suspected to affect the genesis and development of SLE.

A typical SLE microbiota presents with a lower *Firmicutes*: *Bacteroidetes* ratio. A decrease of the families *Christensenellaceae*, *Lactobacillaceae*, *Odoribacteraceae*, *Peptococcaceae*, and *Rikenellaceae* and an increase in family of *Lachnospiraceae* was observed in some studies. Also, several genera like *Streptococcus*, *Campylobacter*, *Veillonella*, *Prevotella*, *Rhodococcus*, *Eggerthelia*, *Klebsiella*, *Eubacterium*, and *Flavonifractor* were abundant in SLE patients (Bankole et al., 2017). From the family *Lachnospiraceae*, the bacteria species *Ruminococcus gnavus* is more prevalent in SLE and could serve as a marker in SLE diagnosis.

2.15.3 Rheumatoid Arthritis (RA)

RA is a chronic autoimmune disease characterized by inflammation and joints pain with various types of systemic involvement. A significant feature of RA is the destruction of bones and cartilage, which may eventually lead to disability. The disease occurs more in women than in men (Xu et al., 2019). The genetic background of the patient plays a vital role in the pathogenesis of the disease. Other risk factors, such as infection and tobacco use, have shown strong evidence for the pathogenesis of the disease. Interaction of metabolites from gut microbiota with the host immune system also leads to the pathogenesis of RA.

Patients who have RA show a decreased diversity in gut microbiota, contrary to the increased diversity exhibited by healthy individuals. Compared to healthy individuals, the gut microbiota of RA patients features a reduced or an increased abundance of different microbial groups. Chen et al. (2016), in their study, found *Eggerthella*, *Faecalibacterium*, and *Colinsella* in fecal samples of patients affected by RA. Increased counts of some *Lactobacilli* species like *L. iners*, *L. ruminis*, and *L. salivarius* were found in both the control group and arthritis patients, while *L. mucosae* were found solely in RA patients.

Different species of the same gut microbiota may play separate roles in the pathogenesis of RA. For example, the abundance of *Prevotella copri* increased in some patients with new-onset RA, but *Prevotella histicola* from the human gut microbiota suppresses the development of RA (Maeda & Takeda, 2017).

2.15.4 Inflammatory Bowel Disease (IBD)

IBD is a chronic inflammatory disorder of the gastro-intestinal tract (GIT) of humans. It is currently a global public health concern. IBD presents in two primary clinical forms: (1) ulcerative colitis (UC) and (2) Crohn's disease (CD), both of which are differentiated by their different clinical forms (Mulder et al., 2014). The hallmarks of IBD include features like disruption in the gut mucosa structure, dysbiosis of gut microbial composition as well as systemic biochemical abnormalities. Inflammatory bowel disease results from gut microbiota dysbiosis, to the extent that a single bacterium that newly colonizes the gut can alter the entire microbiota and lead to the onset of IBD or worsen an already existing IBD condition.

A symbiotic relationship exists between the gut microbiome and the host, whereby the host harbors and feeds the microbiome, which in turn aids the host by performing some physiological functions that keep the host in a healthy state (Israr et al., 2019).

A change in the healthy gut microbiome composition is the main factor responsible for several gastro-intestinal diseases. Thus, the pathogenesis of IBD is associated with gut dysbiosis resulting from the changed interactions between the gut microbiota and the immune system of the mucosa. In terms of microbial load and diversity, there was a significant difference in the gut microbiome of healthy individuals as against IBD patients. The IBD patients have marked reduced commensal bacterial diversity, especially the *Firmicutes* and *Bacteroidetes* phyla, and an increase in the bacteria of the family *Enterobacteriaceae* (Li et al., 2014).

The increase or decrease of specific bacterial groups causes IBD severity. For example, increased prevalence of bacteria belonging to phyla *Actinobacteria* and *Proteobacteria* and decreased prevalence of those in the phylum *Firmicutes* led to disease severity in IBD patients (Zhou et al., 2018). With this, it is very likely that if the gut microbiota gets restored to improve the richness and diversity with beneficial bacteria, it will help in the treatment of IBD patients. It is gladdening to note that Urolithin, a metabolite of gut microorganisms, was linked to a reduction of inflammation and the prevention of colitis (Singh et al., 2019).

2.16 MICROBIOME AND THE ROLE OF PROBIOTICS

Probiotics are living organisms, which, when administered to a host, can exert beneficial health effects by improving gut microbial balance. Probiotics play an essential role in promoting and alleviating infectious diseases. They are mostly harmless for human ingestion with limited reported cases of adverse events. There has been a long history of human consumption of probiotics even without attaching any importance to it, such as the case of using lactic acid bacteria and *bifidobacteria* in fermenting foods and drinks. The most common bacteria found in probiotic products include *Lactobacilli*, *Bifidobacteria*, *Streptococci*, strains of genera like *Bacillus*, *Enterococcus*, *Escherichia*, *Propionibacterium*, and the fungus of genera *Saccharomyces*.

Currently, people use probiotics as a complementary treatment for the improvement of some gastro-intestinal disorders, metabolic syndromes, and vaginal infections (Anukam et al., 2020). A new technique used to treat patients with changes in the gut microbiota is to replace the composition of the abnormal microbiota with a gut microbiota obtained from a donor with a healthy microbiota. This technique is called fecal microbiota transplantation (FMT), and it has been used successfully for some metabolic disorders as well as a variety of gut abnormality like the severe relapsing *Clostridium difficile* infection (Berg et al., 2015a). Other cases where probiotics have been effectively used for the treatment of human health conditions are discussed below.

2.16.1 Vaginal Infection

In a Nigerian probiotics study, seven women diagnosed with BV by Nugent score (7–10) were recruited to provide vaginal and gut sample before and after 14 days oral intake of 3 grams (2.5 × 108 cfu/g) of *Lactobacillus pentosus* KCA1 (Winclove probiotics, Amsterdam) suspended in lukewarm dairy, taken daily for 14 days and re-examined using the Nugent scoring technique. Two out of the women showed evidence of reduced BV by the presence of lactobacilli observed from their vaginal smears and with the Nugent score dropping from 8 to 4. The result showed that the normal vaginal lactobacilli had been gradually replaced, thus inferring the possibility of using probiotics in the treatment of BV (Anukam et al., 2020).

2.16.2 Metabolic Syndrome

In a Belgian study, Cani (2018) gave both live and pasteurized bacteria, *Akkermansia muciniphila* (known to be abundant in the gut of lean people), to insulin-resistant and obese subjects daily for 3 months and reported a decreased cholesterol level and insulin resistance as well as lowered circulating insulin. Probiotics can also be combined with prebiotics (non-digestible polysaccharides or oligosaccharides) to produce a more beneficial effect on a host. This combination called "synbiont" can provide even more benefits than probiotics or prebiotics alone.

2.17 INNOVATIVE APPLICATIONS OF HUMAN MICROBIOME IN MEDICINE

Knowledge about the complex and complicated human microbiome has outpaced all other discoveries in the life sciences in the past decade. Yet the interface of the human microbiome and health is so vast that all the new advances represent only the early stages of a dramatic potential scientific revolution in history. As presented in foregoing sections, there is evidence that dysbiosis of certain microbiome structures play key roles in malnutrition, obesity, and in modulating response of individuals to diet and disease. It is also becoming increasingly clear that beyond the gut, the human microbiome likely impacts all organs through the immune, circulatory, and nervous systems, affecting cognition, behavior, and mental health. Most of these understanding come from correlations and deductions made from genome-based studies using next-generation sequencing. However, to effectively and adequately develop innovative applications of the human microbiome in medicine and other fields, it is critical to overcome many of the current challenges in the approach. So, although there have been significant advances in obtaining microbial genomic data, serious challenges hinder portability and predictability of sequence data. Ability to decode the functional relevance and or redundancies of members of niche consortia is severely limited by lack of direct evidence for causality in complex networks of microbial interactions. Nevertheless, there are growing trends of innovation which over time should empower greater manipulations of microbiomes of the skin, mouth, gut, and beyond with predictable outcomes. A few examples are discussed in this section. The future of personalized medicine lies not in the old methodology of eradication of microbes, but rather individualized microbiome sequencing. Efforts are underway to allow individuals to have their genome sequenced at relatively low cost. However, that alone does not allow for a full picture of what is happening since human genes constitute a mere 1% of gene required for human health. Genomic sequencing when accompanied by microbiome sequencing would provide a deeper appreciation of health and disease, and open avenues to treat disorders. It is probably correct to predict that science is rapidly moving past the age of widespread use of antimicrobials as the primary means to treat infections. Infections and diseases needed to be viewed through the lenses of niches where an infection represents a microbe overtaking a niche at the detriment of others and its host, and whose treatment is the use of beneficial bacteria to reclaim overrun body ecosystems.

2.18 SKIN MICROBIOME APPLICATIONS

2.18.1 Forensic Applications

The human skin is considered one of the largest human organs where its microbiomes are intricately involved in immune regulation and homeostasis of the host among other functions. The diversity and richness of the microbiota that carpet the skin vary significantly at three micro-biotopes that make up different areas of our skin—the sebaceous area (forehead and back), the moist area (nares, inguinal crease, umbilicus, toes, armpit), and the dry area (forearm, palm, buttocks). Although the alpha and beta diversity of the skin microbiomes of young people in a US study conducted by the Esiobu lab showed significant overlap, the marked variations in relative abundances of strains were sufficient to track personal belongings (phones, Tablets, and laptops) of the subjects with up to 80% confidence level, underscoring the potential application of skin microbiomes in forensic fingerprinting. The application of skin microbiome in human forensics is an important emerging technology. Schmedes et al. (2017) reported a novel approach to classify skin microbiomes to their donors by comparing two feature types: *Propionibacterium acnes* pangenome presence/absence features and nucleotide diversities of stable clade-specific markers. They employed supervised learning to attribute skin microbiomes from 14 skin body sites of 12 healthy individuals with accuracies of up to 100% for three body sites, leading to the identification of a subset of 187 markers from 12 clades from each body site that are highly individualizing. Progress in the forensic use of microbiomes requires research into which skin niche is optimally idiotypic and which biomarkers are most suited for distinguishing humans and tracking criminals.

2.18.2 Therapeutic Applications

There is a plethora of models using modulation of the microbiome in acute and chronic wound healing and tissue regeneration. Chronic wounds such as diabetic foot ulcers, decubitus ulcers, and venous leg ulcers are long-term wounds that are particularly hard to treat and heal. One of the major treatment complications in traditional clinical settings is the development of biofilms which add a stunning level of complexity to the treatment, usually with poor prognosis. Biofilms foster pathogenic microbial growth and studies have demonstrated that the formation of a biofilm on a wound site inhibits the coordinated healing process that would otherwise occur. In chronic wound-based biofilms, the genus *Pseudomonas* and *Staphylococcus* are the dominate culprits. In such biofilm bacteria, quorum sensing—a gene expression regulation which coordinates transcriptional output in response to changes in cell population density—is important to survival and expression of virulence.

Targeting the quorum-sensing ability of species represents an emerging application of microbial community research (Conway et al., 2012).

Another innovation with respect to wound healing is the use of the predator bacterium called *Bdellovibrio bacteriovorus* to inhibit the ability of biofilms to preform critical quorum in wounds. *Bdellovibrio bacteriovorus* is a Gram-negative flagellated obligate aerobic bacterium which has a unique life cycle. In its free-living mobile phase, it actively uses chemotaxis to seek out other Gram-negative bacteria. During these mobile phases, metabolic activity is devoted entirely to predation. When *Bdellovibrio bacteriovorus* encounters a prey Gram-negative bacterium, it enters an attack phase, attaching itself to the exterior of its prey using hook-like appendages. It then secrets enzymes that create holes in the outer membrane, allowing it to enter its prey, shed its flagellum as it seals the hole it created. *B. bacteriovorus* then degrades the inner membrane of its prey allowing it access to the cytoplasm on which it feeds. The prey bacterium becomes a bdelloplast. It uses the macromolecules of its host to grow, replicate its DNA, and become a long filament which then fragments into individual mobile progenies which will lyse the remaining bdelloplast (Harini et al., 2013). The energy efficiency of *Bdellovibrio bacteriovorus* can be 0.8–1.5 times more efficient than bacteria growing in rich media. As such, biofilms could prove an ideal hunting ground for the predatory *Bdellovibrio bacteriovorus* whose survival requires a high prey density—one of the hallmarks of biofilms. Harini et al. (2013) reported that *Bdellovibrio bacteriovorus* reduced an existing *Escherichia coli* biofilm biomass by 87% within 3 hours and after a 24-hours exposure period by a 4-log reduction in biofilm cell viability.

Treatment of acute wounds using *Lactobacillus plantarum* is attracting some attention as well. *L. plantarum* is a Gram-positive aerotolerant bacterial probiotic. It has significant antioxidant activities and has already been implement in treatment of gastro-intestinal disorders. In addition, *Lactobacillus plantarum* has been found to inhibit the production of elastase, biofilms, and acyl homoserine lactone produced by *Pseudomonas* in mouse models. *Lactobacillus* also reduced collagen accumulation in *Pseudomonas*-infected burn wounds, leading to a decrease in the appearance of scarring. This therapeutic method has the potential to save lives and reduce the appearance of resulting burn scars (Gonzalez et al., 2016). While more research is needed to define modes of application, these bacteria represent exciting new age of biofilm and wound treatment that shatters the old paradigm of using ineffective antibiotics. Using microbes to treat microbes is a fascinating and novel approach that needs more serious consideration.

2.18.3 Cosmetics and Anti-aging

Although environmental and genetic factors play important roles in aging and senescence, there is growing evidence that the skin and gut microbiome are equally important players. Several budding cosmetic applications employ *Staphylococcus epidermidis* and specific strain of *Nitrosomonas* spp. to improve skin health and reverse signs of aging. Current research include looking at a topical products that contain prebiotics, identifying strains that improve hydration and elasticity of skin, and whose activities are well defined in a changing milieu like the skin. There are also some studies into the protective role of specific strains of *Staph epidermidis* in radiation-induced skin cancers (Nakatsuji et al., 2018).

2.18.4 Oral Microbiomes Applications

The human oral cavity has the second most abundant microbiota in the human body after the GIT, with as many effects on overall human well-being. In 2010, the National Institute of Dental and Craniofacial Research launched a distinct database known as HOMD, the Human Oral Microbiome Database (www.HOMD.org). HOMD was created to maintain the information of oral-derived cultivable and non-cultivable isolates. HOMD has been renamed as eHOMD, expanded HOMD, which provides information of bacterial communities present in the oral cavity, pharynx, nasal passages, sinuses, and esophagus. This type of information will speed up the numerous potential applications of the oral microbes. In a study which compared the oral microbiome of healthy subjects in Nigeria and the United States, we found a core microbiome from the phylum *Firmicutes* amid highly variable consortia. The group *Neisseria* were highly prominent in the Nigerian subjects. In another study, Shoemark and Allen (2014) showed a link between oral bacteria dysbiosis, aging, and Alzheimer's disease. Infection by *P. gingivalis* may result in Alzheimer's because the bacteria produce toxins called gingipains, which are found to accumulate in the brain on Alzheimer's patients. Gingipains degrade human proteins and give rise to misfolded proteins. Similarly, there have been several recent studies observing the effects of the oral microbiome on cardiovascular health, including the effect of nitrate-nitrite-nitric acid pathway and the consequent effects on blood pressure. The nitrate-nitrite-nitric acid pathway relies on oral bacteria to reduce nitrate to nitrite. Nitric oxide regulates vascular endothelial function, therefore regulating blood pressure. Using oral microbes in controlling human blood pressure and heart health represents an important innovative application. Another potential application of the oral microbiomes is in the diagnosis of many diseases whose early detection eluded science before now. Also, Chakraborty and Esiobu (2017) showed that shifts in the community structure of microbial populations of the human mouth is associated with unhealthy habits and disease, and can be explored for developing biomarkers to diagnose diseases. Majority of oral diseases are polymicrobial; therefore, microbiome-based therapies need to be developed. Considering that early colonization of microorganisms has been associated with higher caries incidence, there is a possibility that those bacteria could be used as early diagnostic biomarkers. There is also a chance that these bacteria could provide information for immunization strategies against oral disease. Some species of *Streptococcus* produce high acetaldehyde which is a carcinogen/mutagen substance; therefore, it can be evaluated as a biomarker for diagnosing oral cancer. Gastric cancer is a severe threat to human health; it's the second most common malignancy and one of the leading causes of death in the world. Gastric cancer has a very poor prognosis and as of now there is no routine screening program for it. Early diagnostic and screening methods are very limited, most of the present methods involve invasive procedures such as biopsy, endoscopy, enhanced CT, upper gastro-intestinal angiography, or serology tests. It is crucial that a non-invasive means of early diagnosis for gastric cancer is identified, as such oral biomarkers may be the answer.

2.18.5 Gut Microbiome Applications

The gut microbiome, recently named a human organ presents one of the greatest potentials for innovation and discovery in this century. There is hardly any internal pathways related to human well-being or disease that do not owe to microbial colonization and interaction in and or with the gut. It is well established that more than 1,000 bacterial species, accounting for more than 1 kg of an average human body weight; are involved in immune regulation, food digestion, protection from infection, regulation of the secretion of 90% of the body's happiness hormones (serotonin, dopamine, oxytocin and endorphin levels) as well as direct and indirect regulation of the human brain function and cognition. The latter roles present some of the most intriguing possibilities of manipulating the gut microbiome. The gut microbiota is filled with bacteria of different genera capable of both sensing and producing neurotransmitters that are then able to cross the blood-brain barrier (BBB). More specifically, the presence of an incomplete BBB at the eminence of the hypothalamus and part of the brainstem allows for peripheral circulating factors to gain direct access to the CNS. From the production of dopamine to histamine, over 50 bacterial strains have been identified. Not only do some of these bacteria produce neurotransmitters, but some have also been found to establish quorum sensing via these neurotransmitters, namely norepinephrine. Over ten strains of bacteria have been identified to produce dopamine, with about half of these strains deriving from *Firmicutes*, and some from *Proteobacteria*. Moreover, specific bacteria such as *Bifidobacterium*, *Lactobacillus*, and *Turicibacter* also play roles in synthesis of certain neurotransmitters and proteins essential for a homeostatic microbiome, and consequently, a more stabilized chemical composition in the brain. The overwhelming evidence correlating dysbiotic internal environments with chronic illnesses and disorders that span from IBS to autism to depression should drive home the fact that microbes not only play a role in our systems: they nearly govern the very mechanisms enabling our function. Future research is exploring the increasingly observable projections to different parts of the brain than previously thought, such as the efferent projections of the cerebral cortex, which encompasses the frontal, parietal, and occipital lobes of the brain and thus involve a wider range of neuronal functions serving as vehicles for MGB axis modulation (Berntson et al., 2003). Further reading on this very complex and rapidly growing field can be found in many excellent reviews, including Lozupone et al. (2012). To attempt to include all that is involved in traversing the myriad of pathways guiding microbe-host interactions is to dive down a proverbial rabbit hole with seemingly no return! We shall thus leave the discussion by admiring the movements that neurobiology and microbiology have made in advancing our long-deserved appreciation of the microbial world. While there are many directions that microbes and our neural pathways travel in their highly trafficked functioning, the only direction that future research can go is forward.

2.19 CONCLUSIONS

Plants and animals have evolved intimate relationships with a plenitude of beneficial microbes for varied complementary functions. These microbes form an extension of the host genome and confer phenotypes that the host may not need to develop on their own. Rapid advances achieved in DNA sequencing technologies and computational biology have enhanced the elucidation of various microbiomes to an unprecedented degree, thereby improving our understanding of the task these microbial assemblages perform in shaping crop productivity, resilience to stress and diseases, soil fertility, and human wellness.

Microbiomes have earlier been engineered and deployed to restore microbiota balance for therapeutic purposes in humans. Current research efforts are focusing on the combined application of computer modeling-aided synthetic biology and synthetic ecology as the basis of new living treatments. Lately, the production of synthetic microbial consortia (SMC) and microbiome containing core and hub microbiota have received increased attention owing to their potential to enhance plant fitness to withstand nutritional deficiency as well as environmental stress. Additionally, 3D printing of smart microbes that can recognize and inhibit plant pathogens is well underway for applications on the field. While our knowledge of the various microbiomes cooperatively involved with plants and humans has tremendously improved over the last decade, there is still more to be done in leveraging these microbiome-based technologies in mitigating human health-related issues and the problem of low agricultural productivity. As we anticipate a population explosion in the decades to come, harnessing the benefits of microbiomes could possibly usher in the next generation of medical and agricultural innovations required to sustain man in the foreseeable future.

REFERENCES

Ab Rahman, S. F. S., Singh, E., Pieterse, C. M. J. and Schenk, P. M. (2018). Emerging microbial biocontrol strategies for plant pathogens. *Plant Science*, 267: 102–111.

Agler, M. T., Ruhe, J., Kroll, S., Morhenn, C., Kim, S. T., Weigel, D. and Kemen, E. M. (2016). Microbial hub taxa link host and abiotic factors to plant microbiome variation. *PLoS Biology*, 14(1): e1002352.

Althani, A. A., Marei, H. E., Hamdi, W. S., Nasrallah, G. K., Zowalaty, M. E., Khodor, S. A., Al-Asmakh, M., Abdel-Aziz, H. and Cenciarelli, A. (2016). Human microbiome and its association with health and diseases. *Journal of Cellular Physiology*, 9999: 1–7.

Antharam, V.C., Li, E. C., Ishmael, A., Sharma, A., Mai, V., Rand, K. H. and Wang G. P. (2013). Intestinal dysbiosis and depletion of butyrogenic bacteria in *Clostridium difficile* infection and nosocomial diarrhea. *Journal of Clinical Microbiology*, 51: 2884–2892.

Antony, K. M., Ma, J., Mitchell, K. B., Racusin, D. A., Versalovic, J. and Aagaard, K. (2015). The preterm placental microbiome varies in association with excess maternal gestational weight gain. *American Journal of Obstetrics and Gynaecology*, 212(5): P653.e1–653.e16.

Anukam, K. C., Agbakoba, N. R., Okoli, A. C. and Oguejiofor, C. B. (2019). Vaginal bacteriome of Nigerian women in health and disease: A study with 16S rRNA metagenomics. *Tropical Journal Obstetrics and Gynaecology*, 36: 96–104.

Anukam, K. C., Ejike, C. E., Agbakoba, N. R. and Oguejiofor, C. B. (2020). *Lactobacillus pentosus* KCA1 decreases vaginal and gut microbiota associated with Bacterial Vaginosis (BV), down-regulating IL-1 beta in women of childbearing age and modulates bacterial genes regulated to metabolic functions. (A Preprint). *medRxiv*. https://org/10.1101/2020.04.02.2004023.

Anukam, K. C., Onwunzo, I. N., Olise, N. A., Duru, M. and Agbakoba, N. R. (2018). Oral bacteriome compositions identified by 16S rRNA metagenomics in a randomly selected "healthy" Nigerian male and female subjects. *International Journal of Research and Reports in Dentistry*, 1(1): 1–11.

Bai, J., Jhaney, I. and Wells, J. (2019). Developing a reproducing microbiome data analysis pipeline using the amazon web services cloud for cancer research group: Proof-of-concept study. *JMIR Medical Informatics*, 7(e): e14667.

Bakker, M. G., Manter, D. K., Sheflin, A. M., Weir, T. L. and Vivanco, J. M. (2012). Harnessing the rhizosphere microbiome through plant breeding and agricultural management. *Plant Soil*, 360: 1–13.

Bankole, A., Luo, X. and Husen, Z. (2017). A comparative analysis of gut microbiota between systemic lupus erythematosus patients and non-autoimmune controls: A single center cohort experience. *Lupus Science and Medicine*, 4(Suppl 1): A1–A227.

Bauer, W. D. and Mathesius, U. (2004). Plant responses to bacterial quorum sensing signals. *Current Opinion in Plant Biology*, 7: 429–433.

Belizário, J. E., Faintuch, J. and Garay- Malpartida, M. (2018). Gut microbiome dysbiosis and immunometabolism: New Frontiers for treatment of metabolic diseases. *Mediators of Inflammation*, 2018: 1–12.

Bellabarba, A., Fagorzi, C., DiCenzo G. C., Pini, F., VitI, C. and Checcucci, A. (2019). Deciphering the symbiotic plant microbiome: Translating the most recent discoveries on rhizobia for the improvement of agricultural practices in metal-contaminated and high saline lands. *Agronomy*, 9(9): 529.

Benn, A., Heng, N., Broadbent, J. M. and Thomson, W. M. (2018). Studying the human oral microbiome: Challenges and the evolution of solutions. *Australian Dental Journal*, 63: 14–24.

Berendsen, R. L., Pieterse, C. M. and Bakker, P. A. (2012). The rhizosphere microbiome and plant health. *Trends in Plant Science*, 17(8): 478–486.

Berg, D., Clemente, J. C. and Colombel, J. F. (2015a). Can inflammatory bowel disease be permanently treated with short-term interventions on the microbiome? *Expert Review of Gastroenterology and Hepatology*, 9: 781–795.

Berg, G., Rybakova, D., Grube, M. and köberl, M. (2015b). The plant microbiome explored: Implications for experimental botany. *Journal of Experimental Botany*, 67(4): 995–1002.

Berk, M., Williams, L. J., Jacka, F. N., O'Neil, A., Pasco, J. A., Moylan, S., Allen, N. B., Stuart, A. L., Hayley, Byrne, M. L. and Maes, M. (2013). So depression is an inflammatory disease, but where does the inflammation come from? *BMC Medicine*, 11: 1–16.

Berntson, G. G., Sarter, M. and Cacioppo, J. T. (2003). Ascending visceral regulation of cortical affective information processing. *European Journal of Neuroscience*, 18(8): 2103–2109.

Bhattacharyya, P. N. and Jha, D. K. (2012). Plant growth-promoting rhizobacteria (PGPR): Emergence in agriculture. *World Journal of Microbiology and Biotechnology*, 28: 1327–1350.

Blacher, E., Bashiardes, S., Shapiro, H., Rothschild, D., Mor, U., Dori-Bachash, M., Kleimeyer, C., Moresi, C., Harnik, Y., Zur, M., Zabari, M., Brik, R. B., Kviatcovsky, D., Zmora, N., Cohen, Y., Bar, N., Levi, I., Amar, N., Mehlman, T., Brandis, A., Biton, I., Kuperman, Y., Tsoory, M., Alfahel, L., Harmelin, A., Schwartz, M., Israelson, A., Arike, L., Johansson, M. E. V., Hansson, G. C., Gotkine, M., Segal, E. and Elinav, E. (2019). Potential roles of gut microbiome and metabolites in modulating ALS in mice. *Nature*, 572(7770): 474–480.

Bodenhausen, N., Horton, M. W. and Bergelson, J. (2013). Bacterial communities associated with the leaves and the roots of *Arabidopsis thaliana*. *PLoS One*, 8: e56329.

Bokulich, N. A., Chung, J., Battaglia, T., Henderson, N., Jay, M., Li, H. and Lieber, D. (2016). Antibiotics, birth mode, and diet shape microbiome maturation during early life. *Science Translational Medicine*, 8(343): 343ra82.

Bosch, T.C. and McFall-Ngai, M.J. (2011). Metaorganisms as the new frontier. *Zoology*, 114: 185–190.

Brenner, D., Hiergeist, A., Adis, C., Gessner, A., Albert, C., LudolphJochen, L. and Weishaup, H. (2017). The fecal microbiome of ALS patients. *Neurobiology of Aging*, 61: 132–137.

Bringel, F. and Couée, I. (2015). Pivotal roles of phyllospher microorganisms at the interphase between plant functioning and atmospheric trace gas dynamics. *Frontiers in Microbiology*, 6: 486.

Brown, C.T., Davis-Richardson, A.G., Giongo, A., Gano, K. A., Crabb, D. B., Mukherjee, N., Casella, G., Drew, J. C., Ilonen, J., Knip, M., Hyöty, H., Veijola, R., Simell, O., Neu, J., Wasserfall, C. H., Schatz, D., Atkinson, M. A., Simell, T., and Triplett, E. W. (2011). "Gut microbiome metagenomics analysis suggests a functional model for the development of autoimmunity for type 1 diabetes. *PLoS One*, 6(10): e25792.

Bulgarelli, D., Rott, M., Schlaeppi, K., Ver Loren van Themaat, E., Ahmadinejad, N., Assenza, F., Rauf, P., Huettel, B., Reinhardt, R., Schmelzer, E., Peplies, J., Gloeckner, F. O., Amann, R., Eickhorst, T. and Schulze-Lefert, P. (2012). Revealing structure and assembly cues for *Arabidopsis* root-inhabiting bacterial microbiota. *Nature*, 488: 91–95.

Bulgarelli, D., Schlaeppi, K., Spaepen, S., Ver Loren van Themaat, E. and Schulze-Lefert, P. (2013). Structure and functions of the bacterial microbiota of plants. *Annual Review of Plant Biology*, 64: 807–838.

Cani, P. D. (2018). Human gut microbiome: Hopes, threats and promises. *Gut*, 67(9): 1716–1725.

Chakraborty, S. and Esiobu, N. (2017). Metagenomic rDNA profiles of the oral microbiome in healthy subjects are distinctive compared to that in smokers and disease subjects despite high variations between subjects. *Advances in Microbiology*, 7: 158–174.

Chen, J., Wright, K., Davis, J. M., Jeraldo, P., Marietta, E. V., Murray, J., Nelson, H., Matteson, E. L. and Taneja, V. (2016). An expansion of rare lineage intestinal microbes characterizes rheumatoid arthritis. *Genome Medicine*, 8: 43–92.

Chow, J., Le, S. M., Shen, Y., Khosravi, A. and Mazmanian, S. K. (2010). Host-bacterial symbiosis in health and disease. *Advances in Immunology*, 107: 243–274.

Clemente, J. C., Ursell, L. K., Parfrey, L. W. and Knight, R. (2012). The impact of the gut microbiota on human health: An integrative view. *Cell*, 148: 1258–1270.

Compant, S., Clement, C. and Sessitsch, A. (2010). Plant growth-promoting bacteria in the rhizo- and endosphere of plants: Their role, colonization, mechanisms involved and prospects for utilization. *Soil Biology and Biochemistry*, 42: 669–678.

Compant, S., Samad, A., Faist, H. and Sessitsch, A. (2019). A review on the plant microbiome: Ecology, functions, and emerging trends in microbial application. *Journal of Advanced Research*, 19: 29–37.

Conway, C. A., Esiobu, N. and Lopez, J. V. (2012). Co-Cultures of *Pseudomonas aeruginosa* and *Roseobacter denitrificans* reveal shifts in gene expression levels compared to solo cultures. *The Scientific World Journal*, 120108: 1–5.

Cribby, S., Taylor, M. and Reid, G. (2008). Vaginal microbiota and the use of probiotics. New York. *Interdisciplinary Perspectives in Infectious Diseases*, 8(4): 256–264.

Davies, J. (2001). In a map for human life, count the microbes too. *Science*, 291: 2316.

Dawkins, K. and Esiobu, N. (2017). Arbuscular and ectomycorrhizal fungi associated with the invasive brazilian pepper tree (*Schinus terebinthifolius*) and two native plants in South Florida. *Frontiers in Microbiology*, 8: 665–672.

Dawkins, K. and Esiobu, N. (2018). The invasive Brazilian pepper tree (*Schinus terebinthifolius*) is colonized by a root microbiome enriched with *Alphaproteobacteria* and unclassified *Spartobacteria*. *Frontiers in Microbiology*, 9: 876.

de Goffau, M. C., Lager, S., Sovio, U., Gaccioli, F., Cook, E., Peacock, S. J., Parkhill, J., Charnock-Jones, D. S. and Smith, G. C. (2019). Human placenta has no microbiome but can contain potential pathogens. *Nature*, 572(7769): 329–334.

De Gruttola, A. K., Low, D., Mizoguchi, A. and Mizoguchi, E. (2016). Current understanding of dysbiosis in disease in human and animal models. *Inflammatory Bowel Diseases*, 22: 1137–1150.

De Vrieze, M., Germanier, F., Vuille and Weisskopf, L. (2018). Combining different potato-associated Pseudomonas strains for improved biocontrol of phytophthora infestans. *Frontiers in Microbiology*, 9: 2573.

Dessinioti, C. and Katsambas, A. D. (2010). The role of *Propionibacterium acnes* in acne pathogenesis: Facts and controversies. *Clinics in Dermatology*, 28: 2–7.

Dominguez-Bello, M. G., Godoy-Vitorino, F., Knight, R. and Blaser, M. J. (2019). Role of the microbiome in human development. *Gut*, 68(6): 1108–1114.

Dong, Y. H., Wang, L. H., Xu, J. L., Zhang, H. B., Zhang, X. F. and Zhang, L. H. (2001). Quenching quorum-sensing-dependent bacterial infection by an N-acyl homoserine lactonase. *Nature*, 411(6839): 813–817.

Durán, P., Tortella, G., Viscardi, S., Barra, P. J., Carrión, V. J., Mora, M. L. and Pozo, M. J. (2018). Microbial community composition in take-all suppressive soils. *Frontiers in Microbiology*, 9: 2198.

Escapa, I. F., Chen, T., Huang, Y., Gajare, P., Dewhirst, F. E. and Lemon, K. P. (2018). New insights into human nostril microbiome from the expanded Human Oral Microbiome Database (eHOMD): A resource for the microbiome of the human aerodigestive tract. *mSystems*, 3(6): e00187–18.

Fang, X., Tseng, C-H., Wang, X., Yang, S. G., Meng, F. J., Wang, X. L., Wei, H. and Chen, T. (2016). Evaluation of the microbial diversity in amyotrophic lateral sclerosis using high-throughput sequencing. *Frontiers in Microbiology*, 7: 01479.

Finkel, O. M., Castrillo, G., Paredes, S. H., Gonzalez, I. S. and Dangl, J. (2017). Understanding and exploiting plant beneficial micobes. *Current Opinion in Plant Biology*, 38: 155–163.

Foo, J. L., Ling, H., Lee, Y. S. and Chang, M. W. (2017). Microbiome engineering: Current applications and its future. *Biotechnology Journal*, 12: 1600099.

Gill, S. R., Pop, M., Deboy, R. T., Eckburg, P. B., Turnbaugh, P. J., Samuel, B. S., Gordon, J. I., Relman, D. A., Fraser-Liggett, C. M. and Nelson, K. E. (2006). Metagenomic analysis of the human distal gut microbiome. *Science*, 312: 1355–1359.

Giongo, A., Gano, K. A., Crabb, D. B., Mukherjee, N., Novelo, N. L., Casella, G., Drew, J. C., Ilonen, J., Knip, M., Hyöty, H., Veijola, R., Simell, T., Simell, O., Neu, J., Wasserfall, C. H., Schatz, D., Atkinson, M. A. and Triplett, E. W. (2011). Toward defining the autoimmune microbiome for type 1 diabetes. *The ISME Journal*, 5: 82–91.

Gonzalez, M. R., Fleuchot, B., Lauciello, L., Jafari, P., Applegate, L. A., Raffoul, W., Que, Y. A. and Perron, K. (2016). Effect of human burn wound exudate on *Pseudomonas aeruginosa* virulence. *msphere*, 1(2): e00111–15.

Goodrich, J. K., Waters, J. L., Poole, A. C., Sutter, J. L. Koren, O. and Blekhman, R. (2014). Human genetics shape the gut microbiome. *Cell*, 159: 789–799.

Gopal, M. and Gupta, A. (2016). Microbiome selection could spur next-generation plant breeding strategies. *Frontiers in Microbiology*, 7: 1971.

Gopal, M., Gupta, A. and Thomas, G. V. (2013). Bespoke microbiome therapy to manage plant diseases. *Frontiers in Microbiology*, 4: 355.

Griffen, A. L., Beall, C. J., Campbell, J. H., Firestone, N. D., Kumar, P. S., Yang, Z. K., Podar, M. and Leys, E. J. (2012). Distinct and complex bacterial profiles in human periodontitis and health revealed by 16S pyrosequencing. *The ISME Journal*, 6: 1176–1185.

Gundel, P. E., Rudgers, J. A. and Ghersa, C. M. (2011). Incorporating the process of vertical transmission into understanding of host-symbiont dynamics. *Oikos*, 20: 1121–1128.

Hacquard, S. (2018). Dissecting the role of multi-kingdom microbial consortia on plant health. *The Microbiome: Human and Agriculture in a Microbial World-Conference Report*, The Royal Society, London, October 5, p. 20. Retrieved from royalsociety.org/tof-microbiome.

Hao, D. C. and Xiao, P. G. (2017). Rhizosphere microbiota and microbiome of medicinal plants: From molecular biology to omics approaches. *Chinese Herbal Medicines*, 9(3): 199–217.

Harini, K., Ajila, V. and Hegde, S. (2013). Bdellovibrio bacteriovorus: A future antimicrobial agent? *Journal of Indian Society of Periodontology*, 17(6): 823.

Hirsch, P. (2018). Microbiomes for improving crop production. *The Microbiome: Human Medicine and Agriculture in a Microbial World-Conference Report*, Royal Society, London, October 5, p. 17. Retrieved from royalsociety.org/tof-microbiome.

Israr, K., Naeem, U., Lajia, Z., Yanrui, B., Ashiq, K., Tang Z., Tuanjie, C. and Chunjiang, Z. (2019). Alteration of gut microbiota in inflammatory bowel disease (IBD): Cause or consequence? IBD Treatment Targeting the Gut Microbiome. *Pathogens*, 8(3): 126–164.

JNPD Research (2017). Parkinson's disease linked to gut. Microbiomehttps://www.neurodegenerationresearch.eu/nl/2017/03/parkinsons-disease-linked-to-gut-microbiome/, Accessed on May, 16th 2020.

Keshavarzian, A., Green, S. J., Engen, P. A., Voigt, R. M., Naqib, A., Forsyth, C. B., Mutlu, E. and Shannon, K. M. (2015). Colonic bacterial composition in Parkinson's disease. *Movement Disorders*, 30: 1351–1360.

Kilian, M., Chapple, I. L., Hannig, M., Marsh, P. D., Meuric, V., Pedersen, A. M., Tonetti, M. S., Wade, W. G. and Zaura, E. (2016). The oral microbiome – An update for oral healthcare professionals. *British Dental Journal*, 221: 657–666.

Kim, S., Kwon, S. H., Kam, T. I., Panicker, N., Karuppagounder, S. S., Lee, S., Lee, J. H., Kim, W. R., Kook, M., Foss, C. A., Shen, C., Lee, H., Kulkarni, S., Pasricha, P. J., Lee, G., Pomper, M. G., Dawson, V. L., Dawson, T. M. and Ko, H. S. (2019). Transneuronal propagation of pathologic α-synuclein from the gut to the brain models Parkinson's disease. *Neuron*, 103(4): 627–641.

Korenblum, E., Dong, Y., Szymanski, J., Panda, S., Jozwiak, A., Massalha, H., Meir, S., Rogachev, I. and Aharoni, A. (2020). Rhizosphere microbiome mediates systemic root metabolite exudation by root-to-root signaling. *Proceedings of the National Academy of Science USA*, 117(7): 3874–3883.

Landman, C. and Quevrain, E. (2016). Gut microbiota: Description, role and pathophysiologic implications. *La Revue de Médecine Interne*, 7: 418–423.

Larsen, N., Vogensen, F. K., van den Berg, F.W.J., Nielsen, D. S., Andreasen, A. S., Pedersen, B. K., Al-Soud, W. A., Sørensen, S. J., Hansen, L. H., and Jakobsen, M (2010). Gut microbiota in human adults with type 2 diabetes differs from non-diabetic adults. *PLoS ONE*, 5(2): 1–3, e9085.

Lebeis, S. L., Rott, M., Dangl, J. L. and Schulze-Lefert, P. (2012). Culturing a plant microbiome community at the cross-Rhodes. *New Phytologist*, 196: 341–344.

Li, J., Butcher, J., Mack, D. and Stintzi, A. (2014). Functional impacts of the intestinal microbiome in the pathogenesis of inflammatory bowel disease. *Inflammatory Bowel Diseases*, 21: 139–153.

Li, W., Wu, X., Hu, X., Wang, T., Liang, S., Duan, Y., Jin, F. and Qin, B. (2017). Structural changes of gut microbiota in Parkinson's disease and its correlation with clinical features. *Science China Life Sciences*, 60: 1223–1233.

Lindow, S. E. and Brandl, M. T. (2003). Microbiology of the phyllosphere. *Applied and Environmental Microbiology*, 69: 1875–1883.

Livengood, C. H. (2009). Bacterial vaginosis: An overview. *Reviews in Obstetrics and Gynaecology*, 2(1): 28–37.

Lobell, D. B. and Field, C. B. (2007). Global scale climate–crop yield relationships and the impacts of recent warming. *Environmental Research Letters*, 2: 014002.

Loper, J. E., Hassan, K. A., Mavrodi, D. V. and Davis, E. W. (2012). Comparative genomics of plant-associated Pseudomonas spp.: Insights into diversity and inheritance of traits involved in multitrophic interactions. *PLoS Genetics*, 8(7): e1002784.

Lorenc-Kukula, K., Jafra, S., Oszmianski, J. and Szopa, J. (2005). Ectopic expression of anthocyanin 5-*O*-glucosyltransferase in potato tuber causes increased resistance to bacteria. *Journal of Agricultural and Food Chemistry*, 53: 272–281.

Lozupone, C. A., Stombaugh, J. I., Gordon, J. I., Jansson, J. K. and Knight, R. (2012). Diversity, stability and resilience of the human gut microbiota. *Nature*, 489(7415): 220–230.

Lugtenberg, B. and Kamilova, F. (2009). Plant-growth-promoting rhizobacteria. *Annual Reviews of Microbiology*, 63: 541–556.

Luziatelli, F., Ficca, A. G., Colla, G., Švecová, E. B. and Ruzzi, M. (2019). Foliar application of Vegetal-derived bioactive compounds stimulates the growth of beneficial bacteria and enhances microbiome biodiversity in lettuce. *Frontiers in Plant Science*, 10: 60.

Maeda, Y. and Takeda, K. (2017). Role of gut microbiota in rheumatoid arthritis. *Journal of Clinical Medicine*, 6(6): 60.

Maes, M., Kubera, M., Leunis, J-C., Berk, M., Geffard, M. and Bosmans, E. (2013). In depression, bacterial translocation may drive inflammatory responses, oxidative and nitrosative stress (O&NS), and autoimmune responses directed against O&NS-damaged neoepitopes. *Acta Psychiatrica Scandinavica*, 127(5): 344–354.

Magalhaes, J. G., Tattoli, I. and Girardin, S. E. (2007). The intestinal epithelial barrier: How to distinguish between the microbial flora and pathogens. *Seminars in Immunology*, 19(2):106–115.

Martin, R. (2008). Vaginal microbiota: Composition, protective role, associated pathologies, and therapeutic perspectives. *Enfermedades Infecciosas Microbiología Clínica*, 26(3): 160–167.

Matamoros, S., Gras-Leguen, C., Le Vacon, F., Potel, G. and La Cochetiere, M. (2013). Development of intestinal microbiota in infants and its impact on health. *Trends in Microbiology*, 21(4): 167–173.

Matarazzo, F., Ribeiro, A. C., Feres, M., Faveri, M. and Mayer, M. P. A. (2011). Diversity and quantitative analysis of *Archaea* in aggressive periodontitis and periodontally healthy subjects. *Journal of Clinical Periodontology*, 38: 621–627.

Mauchline, T. H. and Malone, J. G. (2017). Life in earth-the root microbiome to the rescue? *Current Opinion in Microbiology*, 37: 23–28.

Mendes, R., Garbeva, P. and Raaijmakers, J. M. (2013). The rhizosphere microbiome: Significance of plant beneficial, plant pathogenic, and human pathogenic microorganisms. *FEMS Microbiology Reviews*, 37(5): 634–663.

Miquel, S., Martín, R., Rossi, O., Bermúdez-Humarán, L. G., Chatel, J. M., Sokol, H., Thomas, M., Wells, J. M. and Langella, P. (2013). *Faecalibacterium prausnitzii* and human intestinal health. *Current Opinion in Microbiology*, 16: 255–261.

Mmewurum, V. O. (2020). Effect of Probiotics on the axillary microbiome composition of Nigerian males and females. (M.Sc. project – Nnamdi Azikiwe University, Awka, Nigeria).

Mohanram, S. and Kumar, P. (2019). Rhizosphere microbiome: Revisiting the synergy of plant-microbe interactions. *Annals of Microbiology*, 69: 307–320.

Monteiro, R. A., Balsanelli, E., Wassem, R., Marin, A. M., Brusamarello-Santos, L. C. C., Schmidt, M. A., Tadra-Sfeir, M. Z., Pankievicz, V. C. S., Cruz, L. M., Chubatsu, L. S., Pedrosa, F. O. and Souza, E. M. (2013). *Herbaspirillum*-plant interactions: Microscopical, histological and molecular aspects. *Plant Soil*, 356: 175–196.

Mueller, U. G. and Sachs, J. L. (2015). Engineering microbiomes to improve plant and animal health. *Trends in Microbiology*, 23: 606–617.

Mulder, D. J., Noble, A. J., Justinich, C. J. and Duffin, J. M. (2014). A tale of two diseases: The history of inflammatory bowel disease. *Journal of Crohn's Colitis*, 8: 341–348.

Nakatsuji, T., Chen, T. H., Butcher, A. M., Trzoss, L. L., Nam, S. J., Shirakawa, K. T., Zhou, W., Oh, J., Otto, M., Fenical, W. and Gallo, R. L. (2018). A commensal strain of *Staphylococcus epidermidis* protects against skin neoplasia. *Science Advances*, 4(2): eaao4502.

Naseribafrouei, A., Hestad, K., Avershina, E., Sekelja, M., Linløkken, A., Wilson, R. and Rudi, K. (2014). Correlation between the human fecal microbiota and depression. *Neurogastroenterol and Motility*, 26(8): 1155–1162.

Oshim, I.O., Agbakoba, N. R., Oguejiofor, O. C., and Anukam. K. C. (2020). Gut microbiota compositions and modulation of bacterial metabolic functional genes in type -2 diabetes mellitus individuals at Nnewi, Anambra State, Nigeria. *Journal of Medical Laboratory Science*, 30(3): 136–150.

Parasuraman, P., Pattnaik, S. and Busi, S. (2019). *Phyllosphere Microbiome: Functional Importance in Sustainable Agriculture*. Elsevier, Amsterdam, pp. 135–148.

Patil, D. P., Dhotre, D. P., Chavan, S. G., Sultan, A., Jain, D. S., Lanjekar, V. B., Gangawani, J., Shah, P. S., Todkar, J. S., Shah, S., Ranade, D. R., Patole, M. S. and Shouche, Y. S. (2012). Molecular analysis of gut microbiota in obesity among Indian individuals. *Journal of Biosciences*, 37: 647–657.

Pennisi, E. (2019). Evidence mounts that gut bacteria can influence mood, prevent depression. Retrieved from https://www.sciencemag.org/news/2019/02/.

Peters, B. A., Wu, J., Hayes, R. B. and Ahn, J. (2017). The oral fungal mycobiome: Characteristics and relation to periodontitis in a pilot study. *BMC Microbiology*, 17: 157–172.

Pineda, A., Kaplan, I. and Bezemer, T. M. (2017). Steering soil microbiomes to suppress aboveground insect pest. *Trends Plant Science*, 22(9): 770–778.

Proal, A. D., Albert, P. J. and Marshall, T. G. (2014). Inflammatory disease and the human microbiome. *Discovery Medicine*, 17(95): 257–265.

Qiu, Z., Egidi, E., Liu, H., Kaur, S. and Singh, B. K. (2019). New frontiers in agriculture productivity: Optimised microbial inoculants and in situ microbiome engineering. *Biotechnology Advances*, 37: 107371.

Qin, J., Li, Y., Cai, Z, Li, S., Zhu, J., Zhang, F., Liang, S., Zhang, W., Guan, Y., Shen, D., Peng, Y., Zhang, D., and Jie, Z. (2012). A metagenome-wide association study of gut microbiota in type 2 diabetes. *Nature*, 490(7418): 55–60.

Raines, T. (2018). AgBiome: harvesting the plant microbiome. *The Microbiome: Human Medicine and Agriculture in a Microbial World Conference Report*, Royal Society, London, October, 5, p. 19. Retrieved from royalsociety.org/tof-microbiome.

Rajpoot, M., Sharma, A. K., Sharma, A. and Gupta, G. K. (2018). Understanding the microbiome: Emerging biomarkers for exploiting the microbiota for personalized medicine against cancer. *Seminars in Cancer Biology*, 52: 1–8.

Riley, R. (2018). Increasing agricultural productivity: the continuum of breeding, biotech and microbial products. *The Microbiome: Human and Agriculture in a Microbial World-Conference Report*, The Royal Society, London, October 5, p. 10. Retrieved from royalsociety.org/tof-microbiome.

Rogers, G. B., Keating, D. J., Young, R. L., Wong, M. L., Licinio, J. and Wesselingh, S. (2016). From gut dysbiosis to altered brain function and mental illness: Mechanisms and pathways. *Molecular Psychiatry*, 21(6): 738–748.

Rout, M. E. (2014). The plant microbiome. *Advances in Botanical Research*, 69: 279–309.

Rudrappa, T., Czymmek, K. J., Pare, P. W. and Bais, H. P. (2008). Root-secreted malic acid recruits beneficial soil bacteria. *Plant Physiology*, 148: 1547–1556.

Schlaeppi, K. and Bugarelli, D. (2015). The Plant microbe at work. *Molecular Plant- Microbe Interactions*, 28(3): 212–217.

Schmedes, S. E., Woerner, A. E. and Budowle, B. (2017). Forensic human identification using skin microbiomes. *Applied and Environmental Microbiology*, 83(22): e01672–17.

Schulz-Bohm, K., Gerards, S., Hundscheid, M., Melenhorst, J., Boer, W. and Garbeva, P. (2018). Calling from distance: Attraction of soil bacteria by plant root volatiles. *The ISME Journal*, 12: 1252.

Sekar, J., Raj, R. and Prabavathy, V. R. (2016). *Microbial Consortial Products for Sustainable Agriculture: Commercialization and Regulatory Issues in India*. Springer, Singapore, pp. 107–132.

Sessitsch, A., Pfaffenbicher, N. and Mitter, B. (2019). Microbiome applications from lab to field: Facing complexity. *Trends in Plant Science*, 24(3): 194–198.

Shoemark, D. K. and Allen, S. J. (2014). The microbiome and disease: Reviewing the links between the oral microbiome, aging and Alzheimers disease. *Journal of Alzheimers Disease*, 43(3): 725–738.

Singh, R., Chandrashekharappa, S., Bodduluri, S. R., Baby, B. V., Hegde, B., Kotla, N. G., Hiwale, A. A., Saiyed, T., Patel, P., Vijay-Kumar, M., Langille, M. G. I., Douglas, G. M., Cheng, X., Rouchka, E. C., Waigel, S. J., Dryden, G. W., Alatassi, H., Zhang, H. G., Haribabu, B., Vemula, P. K. and Jala, V. R. (2019). Enhancement of the gut barrier integrity by a microbial metabolite through the Nrf2 pathway. *Nature Communications*, 10(1): 89.

Smith, K. P., Handelsman, J. and Goodman, R. M. (1999). Genetic basis in plants for interactions with disease-suppressive bacteria. *Proceedings of the National Academy of Science USA*, 96: 4786–4790.

Sommer, F. and Bäckhed, F. (2013). The gut microbiota-masters of host development and physiology. *Nature Reviews Microbiology*, 11: 227–238.

Stanaway, I. B., Wallace, J. C., Shojaie, A., Griffith, W. C., Hong, S., Wilder, C. S., Green, F. H., Tsai, J., Knight, M., Workman, T., Vigoren, E. M., McLean, J. S., Thompson, B. and Faustman, E. M. (2016). Human oral buccal microbiomes are associated with farmworker status and azinphos-methyl agricultural pesticide exposure. *Applied and Environmental Microbiology*, 83(2): 1–19.

Stiemsma, L. T. and Michels, K. B. (2018). The role of the microbiome in the developmental origins of health and disease. *Pediatrics*, 141(4): e20172437.

Stone, B. W. G., Weingarten, E. A. and Jackson, C. R. (2018). The role of the phyllosphere microbiome in plant health and function. *Annual Plant Reviews*, 1: 1–24.

Stokes, M. E. and McCourt, P. (2014). Towards personalized agriculture: What chemical genomics can bring to plant biotechnology. *Frontiers in Plant Science*, 5: 344.

Trivedi, P., Schenk, P. M., Wallenstein, M. D. and Singh, B. K. (2017). Tiny microbes, big yields: Enhancing food crop production with biological solutions. *Microbial Biotechnology*, 10: 999–1003.

Tseng, C-H. and Wu, C-Y. (2019). The gut microbiome in obesity. *Journal of Formosan Medical Association*, 118(Supplement 1): S3–S9.

Turnbaugh, P. J., Ley, R. E., Hamady, M., Fraser-Liggett, C. M., Knight, R. and Gordon, J. I. (2007). The human microbiome project. *Nature*, 449: 804–810.

Turner, T. R., James, E. K. and Poole, P. S. (2013). The plant microbiome. *Genome Biology*, 14(6): 209.

Valles-Colomer, M., Falony, G., Darzi, Y., Tigchelaar, E. F., Wang, J., Tito, R. Y., Schiweck, C., Kurilshikov, A., Joossens, M., Wijmenga, C., Claes, S., Van Oudenhove, L., Zhernakova, A., Vieira-Silva, S. and Raes, J. (2019). The neuroactive potential of the human gut microbiota in quality of life and depression. *Nature Microbiology*, 4(4):623–632.

Versey, J. K. (2003). Plant growth promoting rhizobacteria as biofertilizers. *Plant Soil*, 255: 571–586.

Vorholt, J. A., Vogel, C., Carlström and Miller, D. B. (2017). Establishing causality: Opportunities of synthetic communities for plant microbiome research. *Cell Host and Microbe*, 22(2): 142–155.

Wegener, C. B. (2001). Transgenic potatoes expressing an *Erwinia* pectate lyase gene – results of a 4-year field experiment. *Potato Research*, 44: 401–410.

Wiest, R. Bacterial translocation (2015). *Bioscience and Microflora*, 24: 61–90.

Xu, H., Liu, M., Cao, J., Li, X., Fan, D., Xia, Y., Lu, X., Li, J., Ju, D. and Zhao, H. (2019). The dynamic interplay between the gut microbiota and autoimmune diseases. *Journal of Immunology Research*, 2019: 7546047.

Xue, C., Penton, C. R., Shen, Z., Zhang, R., Huang, Q., Li, R., Ruan, Y. and Shen, Q. (2015). Manipulating the banana rhizosphere microbiome for biological control of Panama disease. *Scientific Reports*, 5: 11124.

Yap, Y. A. and Marino, E. (2018). An insight into the intestinal web of mucosal immunity, microbiota, and diet in inflammation. *Frontiers in Immunology*, 9: 2617.

Zhai, C. D., Zheng, J. J., An, B. C., Huang, H. F. and Tan, Z. C. (2019). Intestinal microbiota composition in patients with amyotrophic lateral sclerosis: Establishment of bacterial and archaeal communities' analyses. *Chinese Medical Journal*, 132: 1815–1822.

Zhang, H., Liao, X., Sparks, J. B. and Luo, X. M. (2014). Dynamics of gut microbiota in autoimmune lupus. *Applied and Environmental Microbiology*, 80(24): 7551–7560.

Zheng, P., Zeng, B., Zhou, C., Liu, M., Fang, Z., Xu, X., Zeng, L., Chen, J., Fan, S. Du, X., Zhang, X., Yang, D., Yang, Y., Meng, H., Li, W., Melgiri, N. D., Licinio, J., Wei, H. and Xie, P. (2016). Gut microbiome remodeling induces depressive-like behaviors through a pathway mediated by the host's metabolism. *Molecular Psychiatry*, 21: 786–796.

Zhou, J. and Sun, J. (2015). Does the gut drive amyotrophic lateral sclerosis progress? *Neurodegenerative Disease Management*, 5(5): 375–378.

Zhou, Y., Xu, Z. Z., He, Y., Yang, Y., Liu, L., Lin, Q., Nie, Y., Li, M., Zhi, F., Liu, S., Amir, A., González, A., Tripathi, A., Chen, M., Wu, G. D., Rob, K., Zhou, H. and Chen, Y. (2018). Gut microbiota offers universal biomarkers across ethnicity in inflammatory bowel disease diagnosis and infliximab response prediction. *mSystems*, 3: e00188–17.

Zolla, G., Badri, D. V., Bakker, M. G., Manter, D. K. and Vivanco, J. M. (2013). Soil microbiomes vary in their ability to confer drought tolerance to Arabidopsis. *Applied Soil Ecology*, 68: 1–9.

3 The Need for Probiotics in Nigeria

Kingsley C. Anukam
Nnamdi Azikiwe University
Uzobiogene Genomics

Nneka R. Agbakoba
Nnamdi Azikiwe University

CONTENTS

3.1 Introduction 57
3.2 Diarrheal Disease in Children 58
3.3 Westernization of Nigerian Diet 59
3.4 Escalation of Type 2 Diabetes Mellitus (T2DM) in Nigeria 60
3.5 Bacterial Vaginosis 62
3.6 Current Probiotic Market in Nigeria 62
3.7 Conclusion and Recommendation 64
References 64

3.1 INTRODUCTION

A colleague in school asked a very thought-provoking question "Do Africans need probiotics?" "Probiotics should be reserved for the Western world that craves for cleanliness to the extent of losing good bacteria that protects the body against infections and their immunity is now compromised." His question was based on the fact that the use of antiseptics for cleaning is very common in every household and offices. The majority are scared of having babies through the normal vaginal birth and they prefer cesarean section, thereby denying new born babies' access to early microbial colonization. Improved hygiene and better health care reduce the contact of microbial agents to humans, with the negative impact on the immune systems. In some developed countries, many individuals have lost the capacity to fend off some infectious diseases. Some researchers have argued that the increase of asthma, multiple sclerosis, and inflammatory bowel disease may be due to changes in lifestyle for too clean environment. "The hygiene hypothesis" is of the view that the more "hygienic" one becomes, the more one is prone to several autoimmune illnesses.

In Africa, childbirth is mainly through the normal vaginal birth and children play around thereby having direct contact with environmental microbes and thus strong immunity. In this chapter, the objective would be to argue that despite the exposure of various microbes to Africans, especially Nigerians, there are incontrovertible

DOI: 10.1201/9781003180241-3

reasons to consider the health attributes of probiotics. In the year 2001, the Food and Agriculture Organization (FAO) of the United Nations and the World Health Organization (WHO) formed the so-called expert panel without any input from Africans, following the request by the Argentinian government to appraise foods purported to contain probiotics with health and nutritional properties. The panel came up with a definition and probiotics was defined as "live microorganisms which when administered in adequate amounts confer a health benefit on the host." However, in 2014, another self-constituted consensus panel redefined probiotic as: "live microorganisms that, when administered in adequate amounts, confer a health benefit on the host" (Hill et al., 2014). However, the need for probiotics in Nigeria hereby draws credence on the increased hazard of disease and death of some infectious diseases and non-communicable diseases for which probiotics have been found to ameliorate.

3.2 DIARRHEAL DISEASE IN CHILDREN

If exposure to various forms of microbes confers immunity, why do Nigerian children come down with diarrhea leading to high morbidity and mortality? WHO-United Nation's Children Fund (UNICEF) informed in 2009 that diarrhea is the second foremost cause of death among children under five worldwide. Approximately one-in-five child deaths, amounting to about 1.5 million each year, is mainly due to diarrhea disease. It is estimated that diarrhea kills more young children than AIDS, malaria, and measles collectively. UNICEF reported that "only 44% of children with diarrhea in low-income countries receive the recommended treatment." A large number of children are not receiving adequate care for diarrhea in Nigeria (Carvajal-Vélez et al., 2016). Since WHO adopted oral rehydration therapy (ORT) which contains mainly sodium and glucose solution in 1978 as its main means to fight, although the death rate for children suffering from acute diarrhea has drastically fallen from 5 million to 1.3 million deaths yearly, yet one child still dies every 15 seconds, mostly associated with unsafe food and water (https://data.unicef.org/topic/child-health/diarrhoeal-disease/). Nigeria has a newborn death rate of 72.7 deaths/1,000 live births and diarrhea accounts for 16% and this makes it mandatory for the country to continue with the clarion call to meaningfully move forward with the new sustainable development goals (SDGs) by the 2030 target date. Clinical management of acute diarrhea involves replenishment of fluids and electrolytes lost which warrants the use of ORT during the diarrhea episodes in addition to nutritional care. New probiotic formulations can provide fluid and electrolyte replacement in addition to live *Lactobacillus* and or *Bifidobacterium* species that would repopulate the gut thereby producing lactic and acetic acids as metabolic by-products of carbohydrate catabolism. In the last two to three decades, several studies have consistently reported a relatively positive impact of probiotics in treating acute diarrhea in children. For example, products that incorporated *Lactobacillus rhamnosus* GG strain along with *Bifidobacterium bifidum* and *Lactobacillus reuteri* DSM 20016 resulted in the effective treatment of acute bacterial and rotaviral diarrhea without adverse effects (Guandalini et al., 2000; Guarino et al., 1997; Saavedra et al., 1994). In 2008, a review paper reported a statistically significant benefit and moderate clinical benefit of a few, well-identified probiotic strains—mostly *Lactobacillus* GG and *Saccharomyces boulardii*, but also

Lactobacillus reuteri—in the treatment of acute watery diarrhea, primarily rotaviral, in infants (Shornikova et al., 1997) and young children from developed countries (Guandalini, 2008). Although such studies are yet to be carried out in Nigeria, there is no reason to suggest otherwise.

Antibiotic-associated diarrhea (AAD) occurs when the antibiotic alters the microecology of the intestinal microbiota, by causing an imbalance in the diversities and abundance of bacteria in the gut (Keeney et al., 2014). These alterations can affect the capacity of the normal occupant microbiota to resist the invasion of pathogenic microorganisms and their proliferation such as *Clostridium difficile*.

The data on AAD in Nigeria is lacking, but it is reported to stand at 11% in Indian children younger than 2 years. The occurrence of community-acquired *Clostridium difficile* infection (CDI) is increasing in the pediatric population, and among all AAD cases in adult populations, 10%–20% are associated with CDI (Alam and Mushtaq, 2009). A recent Cochrane meta-analysis review that involved 33 studies with 6,352 children reported 8% incidence of AAD after follow-up (5 days to 12 weeks) in those receiving probiotics compared with 19% among those receiving placebo (RR = 0.45, 95% CI 0.36–0.56; number needed to benefit of 9, 95% CI 7–13), thus suggesting an important role of probiotics in preventing AAD in children (Guo et al., 2019) and acute diarrhea (Aggarwal et al., 2014). Agamennone et al. (2018) presented a practical guideline that notifies healthcare professionals and patients on the availability of probiotic products with established efficacy for the prevention of AAD in the Netherlands.

Currently, few probiotics are used in the treatment of diarrhea in children as well as for the treatment of various forms of colitis in adults. Interestingly, *Saccharomyces boulardii* is marketed as FloraNorm in Nigeria and is used to treat pediatric AAD and acute diarrhea in hospitalized infants. Another product "Novolac" is touted to contain five bacterial species (*Lactobacillus acidophilus*, *Bifidobacterium longum*, *Bifidobacterium bifidum*, *Saccharomyces boulardii*, and *Streptococcus thermophilus*) and is claimed for treating diarrhea without any published evidence of efficacy.

3.3 WESTERNIZATION OF NIGERIAN DIET

In the past four decades, Nigeria has witnessed an unprecedented proliferation of fast-food business in urban cities mostly due to the influence of Western society and newfound wealth based on oil revenue. Most families in urban cities order fast foods for breakfast, lunch, and dinner. These foods are mainly high in fat and low in fiber and mostly prepared by frying with high heat which inactivates some vitamins and reduction in nutritional values coupled with the addition of preservatives. The resultant effect is an increase in obesity which alters the body's regulation of energy intake, expenditure, and storage.

Most importantly, food production and consumption has shifted significantly as a result of importation of Western lifestyle ("westernization" of African diets). Majority of imported fast foods are designed with extended shelf-life without taking nutritional needs in mind. Local production of fermented foods seldom imparted to successive young generations. The consequence of these borrowed lifestyle in

Nigeria has led to a significant decrease in the consumption of lactic acid bacteria for the past 60 years. The majority of fast foods are prepared with high-intensity sweeteners such as sucralose, aspartame, and saccharin which have been found to upset the stability and variety of gut microbiota (Nettleton et al., 2016). A review article by Yang et al. (2017) found that consumption of a high-fat diet or Western diet led to widespread fluctuations in gut microbiota of the subjects, thus leading to alterations of food metabolism and nutrient homeostasis. It is noteworthy that before the infiltration of Europeans into Africa, the natives had diets that were rich in plant roots, vegetables, fruits, and tubers that contain prebiotics or non-digestible carbohydrates that promote the growth of lactobacilli and bifidobacterial species. Besides, fermentation processes that introduce lactic acid bacteria and its metabolic products were usually employed to produce and store foods for longer periods (Leach, 2007).

As these lactic acid-producing organisms play a major role in health, it would seem logical to initiate efforts that are aimed at refocusing health restoration and maintenance of the beneficial bacteria in the Nigerian gut. We have already found and determined the genetic make-up of beneficial strain *Lactobacillus pentosus* KCA1 and other bacterial strains that hold great promise, especially for gut and urogenital health, efforts are needed to developed them into various product formulations such probiotic yogurts, probiotic drinks and made available in Nigeria. Recently, our research group has completed the full genome sequence of *Lactobacillus plantarum* strains CNEI-KCA4 (Okpei-Onitsha) (https://www.ncbi.nlm.nih.gov/nuccore/CP053571) and *Lactobacillus plantarum* CNEI-KCA5 (Okpei-Nsukka) (https://www.ncbi.nlm.nih.gov/nuccore/CP059294) isolated from African fermented *Prosopis Africana* seeds.

To counter the impact of fast foods, we suggest that probiotic yogurts could be added as part of the fast-food business which could replenish consumer's gut microbial ecology. In contrast to the commercialization and manufacture of obesogenic food products, the introduction of beneficial microbes in foods has the potential to confer enhancements of health in Nigeria. The market for functional probiotic foods is estimated to reach 74.69 billion dollars by 2025 as forecasted by Fortune Business Insights (www.fortunebusinessinsights.com), but sadly these products are very scarce in Nigeria. Nigeria should take advantage of its microbial diversities and fund research studies that would isolate, test for probiotic properties, carry out clinical trials, and commercialize her microbial products instead of importing ready-made probiotics from the western hemisphere. Bearing in mind the enormous health and economic potential by taking cognizance of the huge population dynamics, translating this area of applied science into the food and healthcare systems will increase economic output and job creation.

3.4 ESCALATION OF TYPE 2 DIABETES MELLITUS (T2DM) IN NIGERIA

Uloko et al. (2018) recently reported an increasing prevalence of diabetes mellitus in Nigeria. There was an increase from 2.2% in 1992 to 5.77% (95% CI 4.3–7.1) in 2018. The prevalence of T2DM in the six geo-political zones of Nigeria were 3.0% north-west, 5.9% north-east, 3.8% north-central zone, 5.5% south-west, 4.6% south-east, and 9.8% south-south zone. Unwholesome dietary custom (8.0%; 95% CI 5.4–10.5) was recorded as the highest risk factor in the prevalence of DM followed by older age (6.6%), urban-dwelling (6.0%), and obesity (5.3%). Several studies have reported dysbiosis, a common denominator of gut microbiota shifting in T2DM, with reduction of butyrate-producing microbes (Qin et al., 2012; Karlsson et al., 2013). The results of our recent study (Oshim et al., 2020) showed depletion in the relative abundance of *Akkermansia* in subjects with T2DM compared with healthy controls as shown in Figure 3.1.

In another comprehensive metagenome wide association study (MGWAS), results showed that patients with T2DM were categorized as having a moderate degree of gut microbial dysbiosis, a reduction in the relative abundance of some known butyrate-producing bacteria, and an increase in some opportunistic pathogens, as well as an enrichment of other microbial functions conferring sulfate reduction and oxidative stress resistance (Qin et al., 2012). Short-chain fatty acids (SCFA) such as butyrate is the main energy source for colonocytes and participates in the maintenance of colonic health. Butyrate-producing bacteria include *Faecalibacterium prausnitzii*, *Roseburia* spp., *Eubacterium hallii*, *Clostridium butyricum*, *Clostridium*

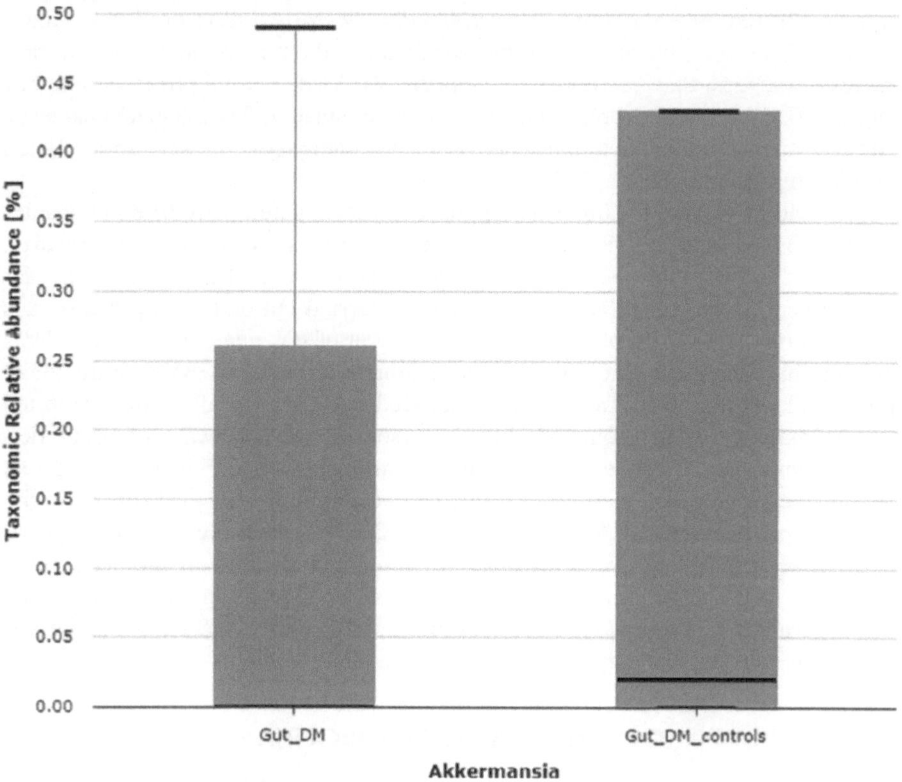

FIGURE 3.1 Relative abundance of *Akkermansia* in the gut of subjects with T2DM and controls without T2DM.

beijerinckii, and *Akkermansia muciniphila* (Louis and Flint, 2009). Oral administration of butyrate significantly decreased blood HbA1c, inflammatory cytokines, inflammatory cell infiltration reduced with a concomitant increase in gut integrity, and intercellular adhesion molecules (Xu et al., 2018). *Akkermansia muciniphila* administration improves glucose profiles and insulin sensitivity (Everard et al., 2013). The consequences that result from the depletion of SCFA production from non-digestible carbohydrates leading to T2DM can be reversed by re-establishing the gut with healthy microbiota that produces SCFA.

3.5 BACTERIAL VAGINOSIS

Bacterial vaginosis (BV) is characterized by alteration of the normal vaginal microbiota from the predominant lactobacilli to a polymicrobial ecology of facultative anaerobes consisting of *Gardnerella*, *Atopobium*, *Corynebacterium*, *Anaerococcus*, *Peptoniphilus*, *Mycoplasma*, *Prevotella*, *Sneathia*, *Eggerthella*, *Mobiluncus*, and *Finegoldia* (Okoli et al., 2019; Ravel et al., 2011). BV affects 14%–50% of reproductive-aged women in Nigeria, and sadly BV is misdiagnosed, inappropriately treated, and underreported. The treatment option is antibiotics, which most times leads to recurrent infections. In our recent study, seven women diagnosed with BV by Nugent score (7–10) were given 3 grams of *Lactobacillus pentosus* KCA1 orally for 5 days, and after 14 days, vaginal and gut sample were analyzed for microbiota compositions. The results suggest that *Lactobacillus pentosus* KCA1 decreased pro-inflammatory cytokine, IL-1 beta, and decreased the relative abundance of BV-associated bacteria, and bacterial metabolic functional genes related to defense systems were up-regulated in the vagina (Anukam et al., 2020).

A myriad of vaginal creams and suppositories formulated with different species of probiotics for vaginal infections are available in drug stores in Nigeria. However, it remained to be ascertained whether they are being prescribed by clinicians for the treatment of vaginal infections. In a survey carried out on health professionals in Nigeria on the use of probiotics for the treatment of BV, none of the respondents agreed to have ever used probiotics in the treatment of recurrent BV as a treatment option (Ejike et al., 2019). In a much earlier study, Anukam et al. (2006c) reported that the majority of the country's health personnel are not aware of probiotics. In a current research carried out on Nigerian women with BV, it was found that some of the women obtained relief after the consumption of yogurt fortified with *Lactobacillus pentosus* KCA1 (Anukam et al., 2013). Studies we conducted previously showed efficacious use of lactobacilli in combination with an antibiotic in the eradication of BV in Nigerian women (Anukam et al., 2006a). In another study, we used intravaginal *Lactobacillus* GR-1 and RC-14 to treat BV subjects, in comparison to standard metronidazole antibiotic gel (Anukam et al., 2006b).

3.6 CURRENT PROBIOTIC MARKET IN NIGERIA

However, Nigeria is replete with reckless and dysfunctional pharmaceutical and biological product control, where anything can be sold to vulnerable people in need. For example, a vaginal cream marketed with a brand name "Klovinal" claims to contain *Lactobacillus* "spores," metronidazole, and clotrimazole. Much as probiotics are said to be harmless, it is good to know that not all species of genera or strains of a species can serve as probiotics. Some species actually may not be good enough to serve as probiotics. Hence there is a need to first screen particular bacteria to be sure that they possess probiotic properties and clinical trials done before approval. The discovery and development of new probiotics are based on results of experiments carried out by researchers to prove that the bacteria species/strain possesses enough beneficial properties for humans. Side effects of probiotics, where they occur, tend to be mild. However, as live organisms, care should be taken when prescribing probiotics to immunocompromised people and those with underlying illnesses.

There is an endless list of microorganisms (bacteria and fungi) used as probiotics and the list continues to increase due to the increase in probiotic research and the increase of newly discovered and identified microbes. Great efforts should be made to prevent abuse of probiotics in Nigeria, unlike antibiotics which is a big problem especially in a country like Nigeria where drug sales are not controlled. All drugs are sold over the counter and nobody is restricted as to what to buy and what not to buy. Even where a doctor rightly prescribes the drugs, patients most often do not stick strictly to the prescribed doses. There is that tendency for patients, especially out-patients, to buy under-dose of the prescribed drugs due to a lack of funds to purchase the complete dose. Even where they get the complete dose, some of them stop taking their drugs once they feel better even if it is 3 or 4 days into a 7-day drug regime. These actions further aggravate the selection pressure in the human system and make way for the emergence of possible new strains of bacteria that will be resistant to the drug when next such a person is exposed to the drug.

Apart from bacteria developing resistance against antibiotics, the antibiotics cause a lot of disturbances on the microbiota of humans. Such dysbiosis is capable of tilting the balance of microbes in favor of harmful bacteria thus leading to disease conditions like antibiotic-associated colitis, infectious diarrhea, vaginal and skin infections, *Helicobacter pylori*, CDIs, autoimmune disorders, and a host of others. Due to the presence of good bacteria, bad bacteria cannot do much harm to the host. So, there must be a continued presence of good bacteria in the system. The best way to keep the bad bacteria under check is by letting good ones compete with them, take their places, and in some cases omit them or decrease them to the minimum safe amount. A good example of this is in the fecal microbiome transplant (FMT). Here good bacteria from healthy gut microbiota are transferred to the intestine of people with colitis where the bad bacteria were out-competed and a healthy microbiota restored. Hence there is a need to always maintain that correct balance to forestall or treat disorders resulting from microbial imbalance; and this is where the need for safer treatment like the use of probiotics comes in. Probiotics come in natural forms and perform many naturally safe activities. However, their maximum use in medicine is just coming up in many developing countries. The use of probiotics in treating and alleviating various disease conditions should be incorporated into the treatment regime in Nigeria especially, to avert the gross antimicrobial abuse that leads to the development of resistant strains. They are generally safe, with virtually no side effects, and most often assist in restoring the body's microbiome to normal. More work needs to be done in the area of enlightening health personnel on the use of probiotics in the replacement of bad microbes with healthy ones as a treatment option of other infections apart from diarrhea.

3.7 CONCLUSION AND RECOMMENDATION

The rise in the importation of probiotic products in Nigeria is an indication of interest shown by healthcare providers and consumers on the importance of beneficial microbes. However, regulatory agencies such as the National Agency for Food and Drug Control (NAFDAC) has to be very vigilant in doing a thorough examination of imported probiotics products on the market. Some of the products may be claiming to be probiotics without appropriate clinical documentation. The internet is inundated with all sorts of probiotics products as there is an increase in people using probiotics for various health reasons or treatment of intestinal, skin, and urogenital disorders. To benefit maximally on probiotics, many efforts should be made to provide a better understanding of the strengths and limitations of probiotic strains.

For Nigeria and indeed Africa to move along with the rest of the world, there should be a sea change in medical and biomedical training curriculum in the universities. Unfortunately, medical training in Nigeria tend to pay more attention to pathogens, and little or no information is given to students on the beneficial microbes that make people healthy as evidenced in Anukam et al. (2006c). We hope that first soon clinicians, especially gastroenterologists, will be armed with lots of information on the rationale for the replenishment of the gut microbiota in dysbiosis and recommend the use of probiotics for specific health benefits more frequently. Second, NAFDAC should establish a branch in her biological, pharmaceutical, food, or natural health product divisions, whose responsibility would be to demand that any imported probiotic products meet FAO/WHO guidelines. Also, probiotic experts should be consulted for enforcement of stringent standards in terms of strain characterization with genomic studies, product formulation, and placebo-controlled clinical evidence of beneficial effects. Third, although Nigeria has a track record for not providing grant funding for functional foods, there is a great need to support the basic understanding of the microbiota that colonizes the Nigerian gut and other body sites. Other nations have already established research consortia such as the "American gut, European gut, Asian gut" to study their microbiome compositions, and in the process, they select species and strains with probiotic properties. Fourth, there is a need to introduce probiotic microbiology in higher institutions where lectures on beneficial microbes need to be integrated into the medical and basic biomedical curricula. Educational information should be made more widely available to people who might be vulnerable to wrong information found on the internet. Everyone should be able to differentiate true probiotics with clinical evidence from products that claim undocumented clinical benefits.

REFERENCES

Agamennone, V., Krul, C.A.M., Rijkers, G., Kort, R. (2018). A practical guide for probiotics applied to the case of antibiotic-associated diarrhea in The Netherlands. *BMC Gastroenterol*, 18, 103. https://doi.org/10.1186/s12876-018-0831-x.

Aggarwal, S., Upadhyay, A., Shah, D., Teotia, N., Agarwal, A., Jaiswal, V. (2014). Lactobacillus GG for treatment of acute childhood diarrhoea: an open labelled, randomized controlled trial. *The Indian Journal of Medical Research*, 139(3), 379–385.

Alam, S., Mushtaq, M. (2009). Antibiotic associated diarrhea in children. *Indian Pediatrics*, 46(6), 491–496.

Anukam, K.C., Ejike, C.E., Agbakoba, N.R., Oguejiofor, C.B. (2020). Lactobacillus pentosus KCA1 decreases vaginal and gut microbiota associated with bacterial vaginosis (BV), down-regulates IL-1 beta in women of child-bearing age and modulates bacterial genes related to metabolic functions. MedRxiv preprint. https://doi.org/10.1101/2020.04.02.20040238.

Anukam, K.C., Macklaim, J.M., Gloor, G.B., Reid, G., Boekhorst, J., Renckens, B., et al. (2013) Genome sequence of Lactobacillus pentosus KCA1: vaginal isolate from a healthy premenopausal woman. *PLoS One*, 8(3), e59239

Anukam, K.C., Osazuwa, E.O., Ahonkhai, I., Ngwu, M., Osemene, G., Bruce, A.W., Reid, G. (2006b). Augmentation of antimicrobial metronidazole therapy of bacterial vaginosis with oral probiotic Lactobacillus rhamnosus GR-1 and Lactobacillus reuteri RC-14: randomized, double-blind, placebo-controlled trial. *Microbes and Infection*, 8(6), 1450–1454.

Anukam, K.C., Osazuwa, E., Osemene, G.I., Ehigiagbe, F., Bruce, A.W., Reid, G. (2006a). Clinical study comparing probiotic Lactobacillus GR-1 and RC-14 with metronidazole vaginal gel to treat symptomatic bacterial vaginosis. *Microbes and Infection*, 8(12–13), 2772–2776.

Anukam, K.C., Osazuwa, E.O., Reid, G. (2006c). Knowledge of probiotics by Nigerian clinicians. *International Journal of Probiotics and Prebiotics*, 1(1), 57–62.

Carvajal-Vélez, L., Amouzou, A., Perin, J., Maïga, A., Tarekegn, H., Akinyemi, A., Shiferaw, S., Young, M., Bryce, J., Newby, H. (2016). Diarrhea management in children under five in sub-Saharan Africa: does the source of care matter? A countdown analysis. *BMC Public Health*, 16, 830. https://doi.org/10.1186/s12889-016-3475-1.

Ejike, C.E., Agbakoba, N.R., Ezeanya, C.C., Anukam, K.C. (2019). Health, social and economic burden of bacterial vaginosis (BV) among Nigerian women of child bearing age: can probiotics restore the vaginal dysbiosis? *Journal Medical Laboratory Science*, 29(2), 37–48.

Everard, A., Belzer, C., Geurts, L., Ouwerkerk, J.P., Druart, C., Bindels, L.B., et al. (2013). Cross-talk between Akkermansiamuciniphila and intestinal epithelium controls diet-induced obesity. *Proceedings of the National Academy of Science USA*, 110, 9066–9071.

FAO/WHO (2001). Evaluation of health and nutritional properties of powder milk and live lactic acid bacteria. Food and Agriculture Organization of the United Nations and World Health Organization Expert Consultation Report. http://www.fao.org/es/ESN/Probio/probio.htm.

Guandalini, S. (2008). Probiotics for children with diarrhea: an update. *Journal of Clinical Gastroenterology*, 42(Suppl 2), S53–S57. doi: 10.1097/MCG.0b013e3181674087.

Guandalini, S., Pensabene, L., Zikri, M.A., Dias, J.A., Casali, L.G., Hoekstra, H., et al. (2000). Lactobacillus GG administered in oral rehydration solution to children with acute diarrhea: a multicenter European trial. *Journal of Pediatric Gastroenterology and Nutrition*, 30, 54–60.

Guarino, A., Canani, R.B., Spagnuolo, M.I., Albano, F., Di Benedetto, L. (1997). Oral bacterial therapy reduces the duration of symptoms and of viral excretion in children with mild diarrhea. *Journal of Pediatric Gastroenterology and Nutrition*, 25(5), 516–519. https://doi.org/10.1097/00005176–199711000–00005.

Guo, Q., Goldenberg, J.Z., Humphrey, C., El Dib, R., Johnston, B. (2019). Probiotics for the prevention of pediatric antibiotic-associated diarrhea. *Cochrane Database of Systematic Reviews*, 4, CD004827.

Hill, C., Guarner, F., Reid, G., Gibson, G.R., Merenstein, D.J., Pot, B., Morelli, L., Canani, R.B., Flint, H.J., Salminen, S., Calder, P.C., Sanders, M.E. (2014). Expert consensus document. The International Scientific Association for Probiotics and Prebiotics consensus statement on the scope and appropriate use of the term probiotic. Nature reviews. *Gastroenterology and hepatology*, 11(8), 506–514. https://doi.org/10.1038/nrgastro.2014.66.

Karlsson, F.H., Tremaroli, V., Nookaew, I., Bergström, G., Behre, C.J., Fagerberg, B., Nielsen, J., Bäckhed, F. (2013). Gut metagenome in European women with normal, impaired and diabetic glucose control. *Nature*, 498(7452), 99–103. https://doi.org/10.1038/nature12198.

Keeney, K.M., Yurist-Doutsch, S., Arrieta, M-C., Finlay, B.B. (2014). Effects of antibiotics on human microbiota and subsequent disease. *Annual Review of Microbiology*, 68, 217–235.

Leach, J.D. (2007). Evolutionary perspective on dietary intake of fibre and colorectal cancer. *European Journal Clinical Nutrition*, 61(1), 140–142.

Louis, P., Flint, H.J. (2009). Diversity, metabolism and microbial ecology of butyrate-producing bacteria from the human large intestine. *FEMS Microbiology Letters*, 294, 1–8.

Nettleton, J.E., Reimer, R.A., Shearer, J. (2016). Reshaping the gut microbiota: impact of low-calorie sweeteners and the link to insulin resistance? *Physiology and Behavior*, 164(Pt B), 488–493. doi: 10.1016/j.physbeh.2016.04.029, pmid:27090230.

Okoli, A.C., Agbakoba, N.R., Ezeanya, C.C., Oguejiofor, C.B., Anukam, K.C. (2019). Comparative abundance and functional biomarkers of the vaginal and gut microbiome of Nigerian women with bacterial vaginosis: a study with 16S rRNA metagenomics. *Journal of Medical Laboratory Science*, 29, 1–26.

Oshim, I.O., Agbakoba, N.R., Oguejiofor, O.C., Anukam, K.C. (2020). Gut microbiota compositions and modulation of bacterial metabolic functional genes in type -2 diabetes mellitus individuals at Nnewi, Anambra State, Nigeria. *Journal Medical Laboratory Science*, 30(3), 136–150. http://doi.org/10.5281/zenodo.4050381.

Qin, J., Li, Y., Cai, Z., Li, S., Zhu, J., Zhang, F., et al. (2012). A metagenome-wide association study of gut microbiota in type 2 diabetes. *Nature*, 490(7418), 55–60. https://doi.org/10.1038/nature11450.

Ravel, J., Gajer, P., Abdo, Z., Schneider, G.M., Koenig, S.S., McCulle, S.L., Karlebach, S., Gorle, R., Russell, J., Tacket, C.O., Brotman, R.M., Davis, C.C., Ault, K., Peralta, L., Forney, L.J. (2011). Vaginal microbiome of reproductive-age women. *Proceedings of the National Academy of Sciences of the United States of America*, 108(Suppl 1), 4680–4687. https://doi.org/10.1073/pnas.1002611107.

Saavedra, J.M., Bauman, N.A., Oung, I., Perman, J.A., Yolken, R.H. (1994). Feeding of Bifidobacterium bifidum and Streptococcus thermophilus to infants in hospital for prevention of diarrhoea and shedding of rotavirus. *Lancet (London, England)*, 344(8929), 1046–1049. https://doi.org/10.1016/s0140-6736(94)91708-6.

Shornikova, A.V., Casas, I.A., Isolauri, E., Mykkänen, H., Vesikari, T. (1997). Lactobacillus reuteri as a therapeutic agent in acute diarrhea in young children. *Journal of Pediatric Gastroenterology and Nutrition*, 24(4), 399–404. https://doi.org/10.1097/00005176–199704000-00008.

Uloko, A.E., Musa, B.M., Ramalan, M.A., Gezawa, I.D., Puepet, F.H., Uloko, A.T., Borodo, M.M., Sada, K.B. (2018). Prevalence and risk factors for diabetes mellitus in Nigeria: a systematic review and meta-analysis. *Diabetes Therapy*, 9(3), 1307–1316. Published online 2018 May 14. doi: 10.1007/s13300-018-0441-1.

Xu, Y.H., Gao, C.L., Guo, H.L., Zhang, W.Q., Huang, W., Tang, S.S., Gan, W.J., Xu, Y., Zhou, H., Zhu, Q. (2018). Sodium butyrate supplementation ameliorates diabetic inflammation in db/db mice. *The Journal of Endocrinology*, 238(3), 231–244. https://doi.org/10.1530/JOE–18-0137.

Yang, B.G., Hur, K.Y., Lee, M.S. (2017). Alterations in gut microbiota and immunity by dietary fat. *Yonsei Medical Journal*, 58(6), 1083–1091. https://doi.org/10.3349/ymj.2017.58.6.1083.

4 Development and Composition of the Human Microbiome from Birth

Toochukwu E. Ogbulie
Federal University of Technology

Nwadiuto (Diuto) Esiobu
Applied Biotech International Nigeria Ltd
Florida Atlantic University

Muinah Fowora
Nigerian Institute of Medical Research (NIMR)

CONTENTS

4.1 Introduction 67
4.2 The Human Skin Microbiome Composition 70
4.3 The Human Urogenital Microbiome 72
4.4 The Microbiome of Oral Cavity 73
4.5 The Human Stomach and Intestinal Microbiome 74
4.6 The Human Lung Microbiome 75
4.7 The Microbiomes and Its Early Development in Human 75
References 77

4.1 INTRODUCTION

The human body harbors a huge array of microorganisms which forma multiple and energetic ecosystems. This diverse ensemble of microorganisms is indeed a complex ecosystem–a social network tightly linked to our health. In the guts of human, zillions of symbiotic microbes made up of primarily bacteria (key player), archaea, fungi and viruses (in certain cases) play huge role in metabolism of essential nutrients in the body. Indeed, they basically cause signal reception by the body especially on the state of hunger and are good in checking the immune system. All the mentioned microorganisms in the gut together with those in the other part of the body constitute the microbiota of human body, whereas the microbiome is the entire

DOI: 10.1201/9781003180241-4

genes contained in the microbiota. However, the two words are interchangeably used. The human microbiome therefore is the assemblage of all the microbiota that reside on or within human tissues (mouth, vagina, uterus, urogenital tract, skin, airways or lung, gut/gastrointestinal tract, mammary glands, placenta, ovarian follicles, conjunctiva, biliary tract) and biofluids (seminal fluid, saliva, sweat, oral mucosa) (Figure 4.1a). A swab sample of these sites as well as an aliquot of the biofluids can give an insight on the diversity of the microbes living on human body.

Generally, mammals are considered as 'super-organisms' with zillions of microbiota which comprise a diverse ecosystem and interact closely with the host. Several intrinsic and extrinsic factors that influence the health of the host, such as the human genome, gender, age and lifestyle and exposed environment, are of critical importance in maintaining the health of the host by creating a dynamic balance between the host and its microbiota. This chapter describes the composition and variation of the microbial communities at different sites in human body with a view to provide an updated overview on the composition of the microbiome in the human from birth.

(a)

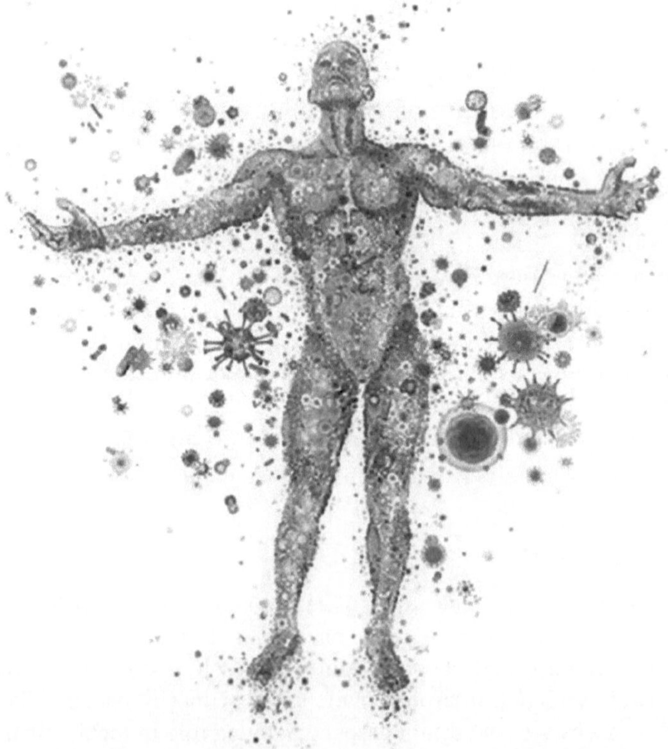

FIGURE 4.1 (a) Our microbiota inhabit all the anatomical sites of our body.

(Continued)

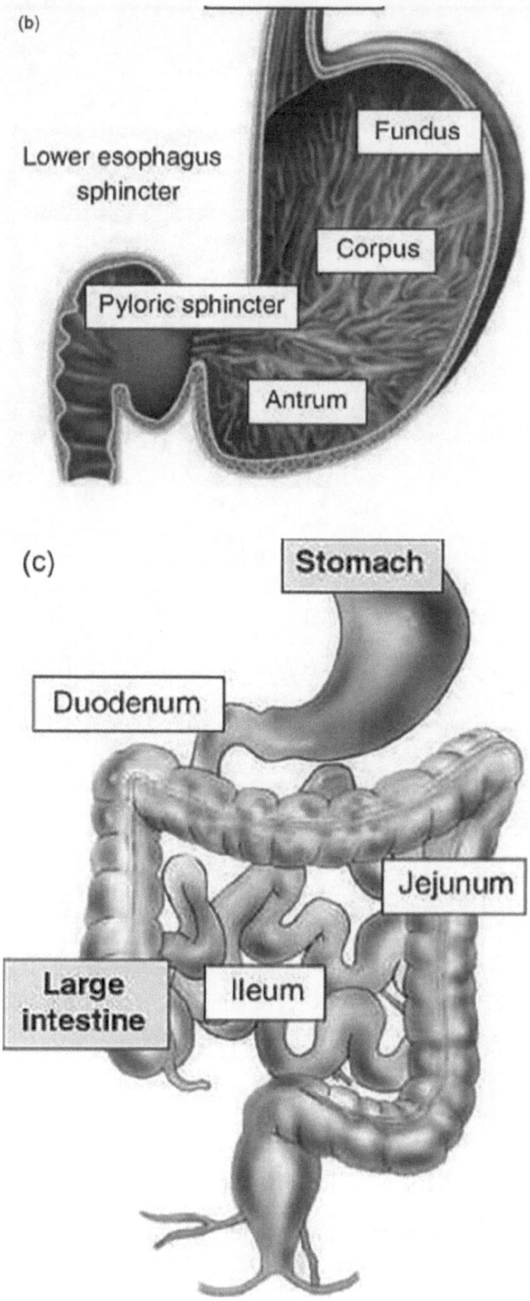

FIGURE 4.1 (*Continued*) (b) The human stomach. (Source: Riedel et al., 2014.) (c) The intestine. (Source: Riedel et al., 2014.)

(*Continued*)

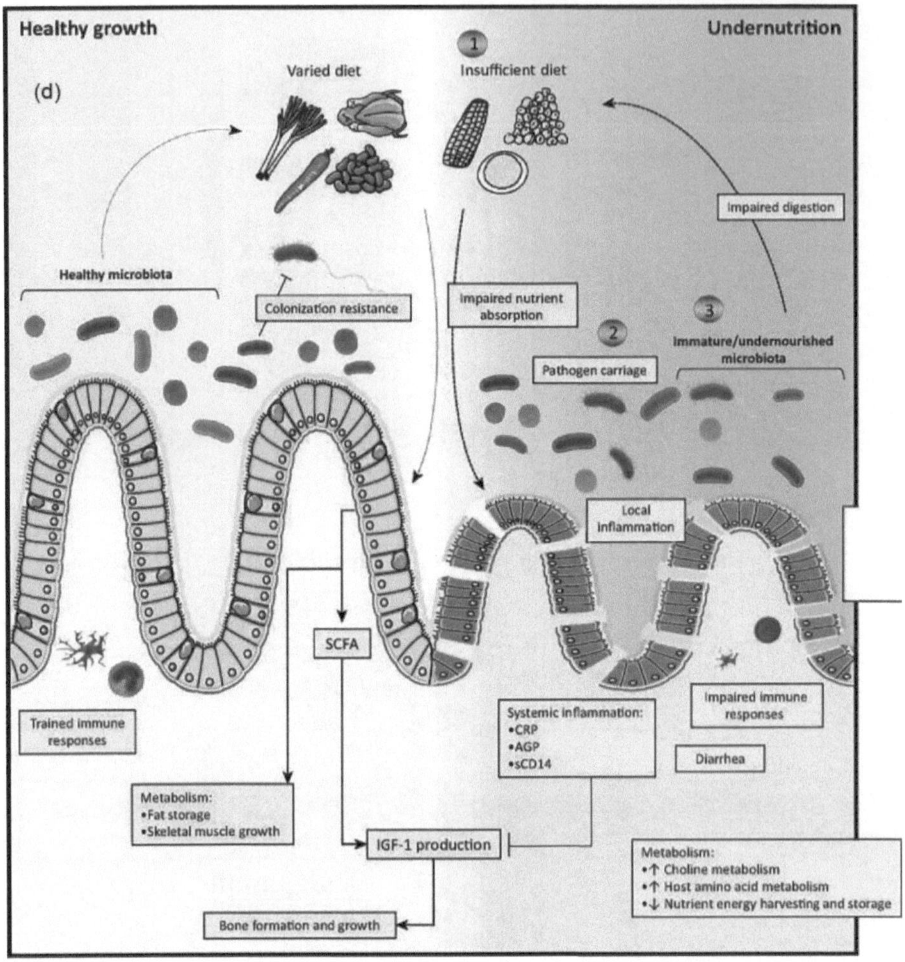

FIGURE 4.1 *(Continued)* (d) The pathways of microbial interaction with host growth in the lumen of healthy versus malnourished children. (Source: Ruairi et al., 2019.)

4.2 THE HUMAN SKIN MICROBIOME COMPOSITION

The skin is made up of the dermis and epidermis layers. It efficiently serves as the body's primary line of defense against infections since microorganisms can only penetrate if there is an opening at the point of landing. Although the skin harbor microbes known as normal body flora, but the surface that is always in contact with microbes is usually acidic; hence, microorganisms with optimum pH > 6 for growth will not proliferate within such environment except those that can adapt to acidic pH.

As the largest organ of the human body, skin harbors diverse microbial communities at different sites with unique niches. The number of bacterial cells on the human skin is estimated to be approximately 10^{12}. The microorganisms form complex interactions within the communities and with the host, influencing the skin health of the host. Occurrences of skin diseases are often accompanied by changes in the composition, structure and function of the skin microbiota.

The skin surface of a neonate is colonized by microorganisms immediately after birth. Most colonized microorganisms are commensals that are highly diverse, including bacteria, eukaryotes and viruses. Most of the microorganisms acquire nutrients from skin secretions, defining an intimate linkage between microbial growth and the physiological state of the skin. Thus, human skin forms a dynamic ecosystem, with microorganisms residing in the unique microenvironment at each site. The micro-environments on the skin vary largely between individuals and within individuals because the structural bases alter considerably under different conditions. For example, genetic diversity results in individual differences in the skin structure. For example, Tao et al. in 2007 studied diversity observing single-nucleotide polymorphisms (SNPs) in the keratin-1 (KRT1) locus. They noted that SNPs in that locus were associated with differences in the migration rate of epidermal keratinocytes. It can also result from different skin sites within each individual since every human display distinct characteristics in their skin structures. Thirdly is the alteration of the skin structure during development and aging.

Furthermore, the human skin microbiota consists of bacteria, eukaryotes and viruses most of which are located in the superficial layers of the epidermis, gland tracts and hair follicles. The composition of the human skin microbiome was made known previously with culture-dependent methods where culture media were used to isolate possible microbiota. But due to the diverse nature of microorganisms and possible presence of culture-independent organisms that may possibly not grow on solid media to form colonies for identification, culture-independent approaches have been found to be more reliable and accurate in determining the true consortium of microbes present in given sample. Many culture-independent methods rely on the polymerase chain reaction (PCR). Highly conserved genes across different taxa, such as the 16S/18S ribosomal RNA (rRNA) gene and certain housekeeping genes such as the rpoB gene and the cpn60 gene as respectively reported by Khamis et al.in 2004 and Chaban et al.in 2009, are often amplified by PCR and used to identify the microbial composition of the microbiota. Denaturing gradient gel electrophoresis (DGGE) was often used in the past to examine microbial diversity based on varied band profiles. When next-generation techniques as such as amplicon, transcriptome and shortgun sequencing are employed, identification of the microorganisms can be made based on the sequence similarity of the sequenced conserved genes to known sequences. 16S rRNA gene sequencing is the method most used in identifying bacterial and archaeal composition in microbial communities. Several 16S rRNA gene databases have been developed and are commonly used in microbiome studies, including RDP, SILVA and Green genes.

Various reports have shown that identification of the biological components of skin microbiota, hence the microbiome using the above culture-independent methods mentioned earlier, revealed the presence of more than 200 bacterial genera residing on the human skin. In 2009, Grice et al. reported that *Actinobacteria*, *Firmicutes* and *Proteobacteria* were the three most prevalent phyla with Actinobacteria being the most dominant bacterial phylum which consists of Gram-positive bacteria with high guanine-cytosine (GC) content, including *Propionibacterium*, *Corynebacterium*, *Mycobacterium*, *Rhodococcus*, *Nocardia* and *Streptomyces*. They further reported that Actinobacteria accounts for 51.8% of the sequences analyzed. However, Costello and colleagues in 2009 showed that 75.4% of the operational taxonomic units (OTUs) found on the forehead of 11 subjects belonged to Actinobacteria. Fitz-Gibbon and his team in 2013 equally observed that *Propionibacteria* are abundant at sebaceous sites, such as the face and upper back, while *Corynebacterium* dominates at the umbilicus, inguinal crease and other moist skin sites. Although bacteria account for the majority of the skin microbiota, other non-bacterial components also colonize the human skin. They include fungi such as the *Malassezia* species which Paulino and colleagues investigated by analyzing its 18S rRNA gene and *Malassezia*-specific 5.8S rRNA/internal transcribed spacer 2 (ITS2); parasites like mites which are small arthropods that frequently colonize the human skin causing human scabies. Study on the skin virome has also been made by Bansal et al. (2002), though it's limited to only a few virus families, such as herpes simplex virus (HSV) and human papillomavirus (HPV). The pathogenicity of these viruses in skin lesions or warts has been widely explored by Barton et al. (2007) and de Jong et al. (2008).

4.3 THE HUMAN UROGENITAL MICROBIOME

As mentioned earlier, the human body is home to a diverse array of microbes that plays an important role in modulating human development, physiology, immunity and nutrition in a mutualistic, relatively site-specific manner; forming communities specific to the various ecological niches lining the body surfaces and cavities. In human urogenitals, the microbial communities just like in other parts of the human body are believed to make up the first line of defense against infection by colonization and competitive exclusion of invasive non-indigenous organisms and opportunistic pathogens native to other body sites. Thus, the female lower reproductive tract is a good environment for microbes – collectively known as the vaginal microbiota. They play critical roles in protecting and maintaining the health of this organ. The protective effect in women at their reproductive age as reported in 2010 by lInhares and co-workers is presumptively due to the production of lactic acid, hydrogen peroxide and other antimicrobial substances. These substances lead to reduction in the pH of the vaginal environment, precluding unwanted microbial growth, making it inhospitable to other invading organisms. High proportions of lactic acid bacteria (LABs), particularly *Lactobacillus* spp., have been reported in 2011 by O'Hanlon et al. to be the distinguishing characteristics of vaginal health lack of which could be typically associated with an abnormal condition. However, despite the prevailing views that *Lactobacillus* is a prerequisite for vaginal health, previous studies in 1990 by Redondo-Lopez et al. using culture-dependent methods have shown a diverse array of other bacteria of genus *Staphylococcus*, *Ureaplasma*, *Corynebacterium*, *Streptococcus*, *Peptostreptococcus*, *Gardnerella*, *Bacteroides*, *Mycoplasma*, *Enterococcus*, *Escherichia*, *Veillonella* and *Bifidobacterium*, as well as the yeast Candida are part of the viginal microbiota.

Furthermore, Onderdonk et al. and Todar in 1986 and 2012respectively illustrated their observed variability in both the composition and ratio of the vaginal microbiota; although their findings showed that it all depends on the menstrual cycle stage. Antonio and his group on the other hand reported that ethnicity have also influenced vaginal flora. Their study on African-American women showed higher vaginal pH and lower number of hydrogen peroxide-producing lactobacilli. In 2007, Witkin and coin their study linked the use of antibiotics and sexual intercourse as influential factors to the lactobacilli loss whereas the use of condom during intercourse increases the population of *Escherichia coli* with variable level of lactobacilli around the vagina. Such changes in the normal, healthy vaginal microflora/microbiota evidently indicate progressive infections like candidiasis and bacterial vaginosis. This situation, however, inhibits the growth of *Lactobacillus* species with the latter producing H_2O_2 that inhibits the growth of *Candida albicans* in a normal healthy woman. Cui and co-workers also observed the presence of other fungal genera in the vagina to include Pichia, Alternaria, Rhodotorula, Eurotium and *Cladosporium*, among others.

Comparatively, the composition of the penile microbiome and its significance to health is poorly understood. Previous reports have shown that the male urogenital tract, especially the urethra, was considered to be sterile or only incidentally colonized by transient microbiota, except in men with sexually transmitted infections (STIs). This view was supported by three basic facts as

i. the relative paucity of urinary tract infections (UTIs) in men compared to women
ii. convergence of the distal urinary and reproductive tracts in men permitting simultaneous washing of openings of both tracts by urine passage and
iii. wide anatomic separation of the urethral and anal orifices in men.

Fethers and coin 2012 suggested that the composition and dynamics of male urogenital tract microbiomes could have ramifications for health and susceptibility to infectious disease in both men and women. Seminal study in 1977 by Bowie et al. using urethral swab, prostatic massage and voided urine specimens from men with and without non-gonococcal urethritis (NGU) was, however, the first to show convincingly that the adult male urethra supported viable microorganisms and that the composition of urethral microbiomes differed in men with and without NGU. One or more aerobic bacteria, including *Staphylococus epidermidis*, *Corynebacterium* spp., *G. vaginalis* and Alpha-haemolytic streptococci, in order of frequency, were cultured from >95% of NGU-positive men and 100% of NGU-negative men. Interestingly, *Lactobacillus*, *Gardnerella*, *Streptococcus*, and Gram-negative anaerobes were more frequent in men who did not have NGU.

4.4 THE MICROBIOME OF ORAL CAVITY

The microbiome of human oral cavity constitutes all the microbiota found on or in the mouth of human and its contiguous extensions (the tonsils and pharynx). Studies on saliva by Curtis and co in 2011 showed that at the normal salivary production rate of 750 mL/day in human, millions of bacterial cells are released. Dewhirst et al. in 2010 isolated and identified many bacterial species from the oral cavity using culture-based method, whereas sequencing with 16S rRNA taxonomical marker showed phylogenetic similarity of about 360 phylotypes. Indeed, the study of Sherwood and his team showed that the environment present in the human mouth provided a source of water, nutrients and moderate temperature for the microbiota permitting the growth of characteristic microorganisms found there. Wang and co in 2014 observed that resident microbes of the buccal cavity usually resist mechanical flushing into the stomach by adherence. Hence, they escape destruction by stomach acidity. Nevertheless, *Lactobacillus*, *Actinomyces*, *Propionibacterium*, *Arachnia*, *Bacteroides*, *Peptostreptococcus*, *Bifidobacterium*, *Eubacterium*, *Fusobacterium*, *Leptotrichia*, *Selenomonas*, *Treponema*, *Peptococcus*, and *Veillonella* and *Cryptococcus Candida*, *Fusarium*, *Aspergillus*, *Cladosporium*, *Glomus*, *Penicillium*, and *Alternaria* are, respectively, the bacterial and fungal flora evidently identified in the mouth.

4.5 THE HUMAN STOMACH AND INTESTINAL MICROBIOME

The human stomach (Figure 4.1b) and the intestine (Figure 4.1c) are the major parts of the gastrointestinal tract (GIT). The stomach is a J-shaped internal organ made of three parts – the fundus (upper part), the corpus (main body) and an antrum (lower part) – connected to the duodenum through the pyloric sphincter.

As a temporary food store, where food are mixed with gastric juice to enhance chiming followed by protein pre-digestion, the stomach harbors consortium of microorganisms that forms a complex microbial ecosystem. They are known as the gut microbiota with density reported to be highest in the colon. They play vital roles in human health maintenance, although intestinal dysbiosis (variation in composition) has been reported to possibly lead to the onset of diseases such as colorectal cancer, bacterial infections, allergic reactions, and metabolic disorder. The pathway in Figure 4.1d is an illustration of the wholesomeness of the structure and function of the intestinal barrier and its mechanism of regulation by the gut microbial composition in healthy and malnourished infants. Ruairi et al. in 2019 depict that a good gut microbiome provides colonization resistance against invading pathogens during healthy growth condition, the interaction which enhances maintenance of immune homeostasis and antigen recognition by the innate immune system. Their report suggested that a microbiome imbalance in early life may impair growth pathways as well as intestinal barrier protection resulting in impaired immune response, blunting of the villous, mucus degradation and permeability in the intestine. This may be a potential contributor to impaired growth and development of a child due to enteric dysfunction, inflammation of the whole body system, infectious diseases and diarrhea.

In other words, impairment in amino acid metabolism which is a key nutrient in human can lead to growth inhibition due to inadequate or insufficient diet (1), carriage of pathogen (2) and/or microbiome imbalance (3). This can possibly induce EED and under-nutrition phenotypes. Figure 4.1d, however, shows that the normal growth hormones production can also be impaired due to composition of the gut microbiota. However, re-establishment of such condition for enhanced normal homeostasis of the gut microbiota can be achieved through the use of biotherapeutic agents such as indigestible food components (prebiotics), as well as the health-promoting viable microorganisms (probiotics).

Indeed most microorganisms cannot survive in the stomach due to the high acidity, molecular diversity studies have shown that Proteobacteria, Firmicutes, Bacteroidetes and Actinobacteria dominates the stomach but the main bacterial inhabitants include *Streptococcus* spp., *Staphylococcus* spp., *Lactobacillus*, *Peptostreptococcus*, *Gemella*, *Actinomyces*, *Prevotella*, *Helicobacter pylori* and varying yeast types.

Comparatively, the microbiota of the mouth and large intestine or feces has been intensively investigated more than small intestine due to the restricted accessibility of the intestine. Hence, there is less data on the small intestinal microbiota of healthy individuals. Nevertheless, only a few microbiota such as acid-tolerant streptococci and lactobacilli have been reported to colonize the mucous membrane of the duodenum and jejunum tubes but at low density. However, toward the end of the ileum, the lumen was dominated by *Streptococci*, *Enterococci* and coliforms. However, the large intestine consists of all microbiota inhabiting the large intestine in the caecum, colon, rectum and anal canal. The use of molecular-based and high-throughput techniques have therefore been employed in assessing the diversity of these gut flora which had proven to be more efficient than culture-based approaches (Lloyd-Price et al., 2016).

4.6 THE HUMAN LUNG MICROBIOME

In the human body, oxygen and carbon dioxide gas exchange occurs in the lungs; hence, its functionality maintenance is critical to human health. Thus, damage to the lungs can result from a wide range of causes which can impair its function representing a major indicator of mortality. Although the upper and lower respiratory systems possess preventive mechanisms against microbes such as removal of microbes from the respiratory system by the use of the goblet cells' mucous to trap microbes through continuous moving of the ciliated epithelial cells and bactericidal effect using lysozyme enzyme by nasal mucus, it still has its own set of microbiota. Reports have shown the presence of the normal flora of the respiratory tract to respectively belong to the following bacterial and fungal genera; *Prevotella* sp., *Sphingomonas* sp., *Pseudomonas* sp., *Acinetobacter* sp., *Fusobacterium* sp., *Megasphaera* sp., *Veillonella* sp., *Staphylococcus aureus*, *Haemophilus influenzae*, *Streptococcus pyogenes*, *Neisseria meningitidis*, *Streptococcus pneumoniae* and *Candida* sp., *Malassezia* sp., *Neosartorya* sp., *Saccharomyces* sp., and *Aspergillus* sp., some of which can cause serious opportunistic infections especially in immunocompromised individuals.

4.7 THE MICROBIOMES AND ITS EARLY DEVELOPMENT IN HUMAN

Human body is an excellent host to zillions of microbes that inhabit it and the complex, body–habitat-specific, adaptive ecosystems they create usually sync to the changing host physiology. In humans, the gastrointestinal microbiome composition is established during the first year of life, whereby the human microbiota dramatically changes through interactions with the developing immune system. Various environmental factors played obvious roles during the process of establishing this microbiota, with the potential to be a main determinant of life-long health. Such factors influence early-life microbiome dynamics and equally serve as important precipitators of a healthy microbiome.

The early development of human microbiome has been reported to ensue during the first 1,000 days of life. This according to Ruairi and his team in 2019indicates that rapid maturation of certain pathways such as endocrine, metabolic, neural and immune pathways influences and supports growth and development of a child during prenatal and early postnatal period. They added that when these pathways which work in tandem are affected by adverse environmental factors like infection or under-feeding, the course of growth and development of the child can be hindered, leading to malnutrition. This, as explained by Mohammadkhah et al. in 2018, can consequently manifest as over-nutrition (overweight or obesity) or as under-nutrition –stunting or wasting.

Furthermore, new-born life actually begins sterile but picks up commensal bacteria from the mother during birth (either by Cesarean section or vaginal delivery) and breastfeeding, and continues to be exposed through contact with parents, grandparents, siblings, pets and the environment itself, which is crowded with bacteria. Studies have recently shown varying other agents that negatively affects the microbiome of a new life to be C-section method of delivery (Turroni et al., 2020). This harbors more pathogenic strains of *Escherichia coli* and *Staphylococcus aureus* leading to delay in development of non-pathogenic/beneficial gut microbiota. Contrary to this report, child delivery through the vaginal canal has the advantage of faster development of more beneficial gut microbiota with negligible difference to that of the mother (Mohammadkhah et al., 2018).Other agents include through insufficient breastfeeding, antibiotic use, an overly hygienic environment, stress and poor diet (Dominguez-Bello et al., 2019).

A typical example was the largest clinical microbiome study in infants reported in 2018 where the sequence of microbial colonization in the infant gut through age 4 had been explored by a team of researchers led by Christopher Stewart in 2018 at Baylor College of Medicine known as TEDDY (The Environmental Determinants of Diabetes in the Young). Their findings on the distinct stages of development in the microbiome in association with exposure in early life were published in the journal *Nature*.

They sampled infants within 3–46 months of age; they analyzed the microbiome using 16S rRNA gene sequencing. The result of their study showed three distinct phases of gut microbiome development to include

1. Phase of development between 3 and 14 months of age
2. Phase of transition within 15–30 months of age
3. Phase of stability within 31–46 months of age.

Their findings clearly indicates that microbiome structure from various study can be influenced by receipt of breast milk, either exclusive or partial, and breastfeeding was associated with higher density of *Bifidobacterium* species such as *B. breve* and *B. bifidum* which have led to so many cases of breast milk cessation and faster maturation of the gut microbiome, predominated by the phylum Firmicutes. On the other hand, density of *Bacteroides* (*Bacteroides fragilis*) was also observed to be significantly high during the developmental phase in vaginally delivered infants and is equally associated with increased gut diversity and faster maturation, regardless of the birth mode.

The report of this team has obviously shown that 0–4 years are the critical years for a child's health. It is the period when their gut microbiome is essentially forming which plays significant role in immune function, digestion, hormone signaling, vitamin production and micro/macronutrient absorption. Adequate nutrient intake is therefore of great importance for the development of a microbiome/microbiota as it is for the early development of our child's brain and heart.

REFERENCES

Antonio, M. A. D., Hawes, S. E. & Hillier, S. L. (1999). The identification of vaginal lactobacillus species and the demographic and microbiologic characteristics of women colonized by these species. *The Journal of Infectious Diseases,180*(6), 1950–1956.

Bansal, R., Tutrone, W. D. & Weinberg, J. M. (2002). Viral skin infections in the elderly: Diagnosis and management. *Drugs and Aging*, *19*, 503–514.

Barton, E. S. et al. (2007). Herpesvirus latency confers symbiotic protection from bacterial infection. *Nature*, *447*, 326–329.

Bowie, W.R., Pollock, H.M., Forsyth, P.S., Floyd, J.F., Alexander, E.R., Wang, S.P., et al. (1977). Bacteriology of the urethra in normal men and men with non-gonococcal urethritis. *Journal of Clinical Microbiology,* 6, 482–488.

Chaban, B., Musil, K. M., Himsworth, C. G. & Hill, J. E. (2009).Development of cpn60-based real-time quantitative PCR assays for the detection of 14 Campylobacter species and application to screening of canine fecal samples. *Applied and Environmental Microbiology*, *75*(9), 3055–3061.doi:10.1128/AEM.00101-09.

Costello, E. K., Lauber, C. L., Hamady, M., Fierer, N., Gordon, J. I. & Knight, R. (2009). Bacterial community variation in human body habitats across space and time. *Science*, *326*, 1694–1697.

Cui, L., Morris, A. & Ghedin, E. (2013).The human mycobiome in health and disease. *Genome Medicine*, *5*(7), 63–67.

Curtis, M.A., Zenobia, C. & Darveau, R.P. (2011). The relationship of the oral microbiota to periodontal health and disease. *Cell Host and Microbe*, 10, 302–306.

De Jong, M. A., De Witte, L., Bolmstedt, A., Van Kooyk, Y. & Geijtenbeek, T. B. (2008). Dendritic cells mediate herpes simplex virus infection and transmission through the C-type lectin DC-SIGN. *Journal of General Virology*, *89*, 2398–2409.

Dewhirst, F.E., Chen, T., Izard, J., Paster, B.J., Tanner, A.C.R., Yu, W.H., et al. (2010). The human oral microbiome. *Journal of Bacteriology*, *192*, 5002–5017.

Dominguez-Bello, M.G., Vitorino, F. G., Knight, R. & Blaser, M. J. (2019).Role of the microbiome in human development. *Recent Advances in Basic Science*, *68*, 1108–1114. doi: 10.1136/gutjnl-2018-317503.

Fethers, K., Twin, J., Fairley, C. K., Fowkes, F. J., Garland, S. M. & Fehler, G. (2012).Bacterial vaginosis (BV) candidate bacteria: Associations with BV and behavioral practices in sexually experienced and inexperienced women. *PLoS ONE*, *7*, e30633.

Fitz-Gibbon, S., Tomida, S., Chiu, B., Nguyen, L., Du, C., Liu, M. & Li, H. (2013). *Propionibacterium acnes* strain populations in the human skin microbiome associated with acne. *Journal of Investigative Dermatology*, *133*(9), 2152–2160.doi: 10.1038/jid.2013.21.

Grice, E. A., Kong, H. H., Conlan, S., Deming, C.B., Davis, J. & Young, A.C. (2009). Topographical and temporal diversity of the human skin microbiome. *Science*, *324*, 1190–1192.

lInhares, I. M., GIraldo, P. C. & Baracat, E. C. (2010). New findings about vaginal bacterial flora. *Revista Da Associacao Medica Brasileira*, *56*(3), 370–374.

Khamis, A., Raoult, D. & La Scola, B. (2004). *rpoB* gene sequencing for identification of *Corynebacterium* species. *Journal of Clinical Microbiology*, *42*(9), 3925–3931. doi: 10.1128/JCM.42.9.3925-3931.

Lloyd-Price, J., Abu-Ali, G. & Huttenhower, C. (2016). The healthy human microbiome. *Genome Medicine*, *8*, 51–61.

Marchesi, J. R. (2010). Prokaryotic and eukaryotic diversity of the human gut. *Advances in Applied Microbiology*, *72*, 43–62.

Mohammadkhah, A. I., Simpson, E. B., Patterson, S. G. & Ferguson, J. F. (2018). Development of the gut microbiome in children, and lifetime implications for obesity and cardiometabolic disease. *Children,* 5(12): 160–180. doi:10.3390/children5120160

O'Hanlon, D. E., Moench, T. R. & Cone, R. A. (2011). In vaginal fluid, bacteria associated with bacterial vaginosis can be suppressed with lactic acid but not hydrogen peroxide. *BMC Infectious Diseases*, *11*, 200–207. doi: 10.1186/1471-2334-11-200.

Onderdonk, A. B., Zamarchi, G. R., Walsh, J. A., Mellor, R. D., Muñoz, A. & Kass, E. H. (1986). Methods for quantitative and qualitative evaluation of vaginal microflora during mensturation. *Applied and Environmental Microbiology*, *51*(2), 333–339.

Redondo-Lopez, V., Cook, R. L. & Sobel, J. D. (1990). Emerging role of lactobacilli in the control and maintenance of the vaginal bacterial microflora. *Review of Infectious Diseases*, *12*, 856–872.

Ruairi, C. R., Amee, R. M., Finlay, B. B. & Prendergast, A. J. (2019).The human microbiome and child growth first 1000days and beyond. *Trends in Microbiology*, *27*(2), 131–147.

Sherwood, L., Willey, J. & Woolverton, C. (2013). *Prescott's Microbiology* (9th ed.). New York: McGraw Hill. pp. 713–721.

Stewart, C. J. et al. (2018). Temporal development of the gut microbiome in early childhood from the TEDDY study. *Nature*, *562*, 583–602. https://doi.org/10.1038/s41586-018-0617.

Tao, H., Berno, A.J., Cox, D. R. & Frazer, K. A. (2007). *In vitro* human keratinocyte migration rates are associated with SNPs in the *KRT1* interval. *PLoS ONE*, *2*(8): e697. https://doi.org/10.1371/journal.pone.0000697.

Todar, K. (2012). The normal bacterial flora of humans. Retrieved April 2020 http://textbookofbacteriology.net/normalflora_3ht.

Turroni, F. et al. (2020). The infant gut microbiome as a microbial organ influencing host well-being. *Italian Journal of Pediatrics*, *46*, 16–29.

Wang, Z. K., Yang, Y. S., Stefka, A. T., Sun, G. & Peng, L. H. (2014). Fungal microbiota and digestive diseases. *Alimentary Pharmacology and Therapeutics*, *39*(8), 751–766.

Witkin, S. S., Linhares, I. M. & Giraldo, P. (2007). Bacterial flora of the female genital tract: Function and immune regulation. *Best Practice & Research Clinical Obstetrics & Gynaecology*, *21*(3), 347–354.

5 Microbiota Transplantation, Health Implications, and the Way Forward

Olugbenga Samuel Michael, Olufemi Idowu Oluranti, and Ayomide Michael Oshinjo
Bowen University

Charles Oluwaseun Adetunji
Edo State University Uzairue

Juliana Bunmi Adetunji
Osun State University

Nwadiuto (Diuto) Esiobu
Applied Biotech International Nigeria Ltd
Florida Atlantic University

CONTENTS

5.1 Introduction 80
5.2 FMT Complications 81
5.3 Therapeutic Uses of FMT 81
5.4 Diet and Gut Microbiota 82
5.5 Donor Screening for FMT 83
5.6 Fecal Microbiota Delivery Methods 85
5.7 Application of FMT 86
5.8 Clostridium Difficile Infection 87
5.9 Inflammatory Bowel Disease 87
5.10 Cardiometabolic Disorders 87
5.11 Vaginal Microbiota Transplantation 87
5.12 Intestinal Microbiota Transplantation 89
5.13 Limitations of the FMT 90
5.14 Future Trends of Microbiota Transplantation 91
5.15 Conclusion 91
References 92

DOI: 10.1201/9781003180241-5

5.1 INTRODUCTION

We live in a microbial world and the human body is in constant contact with microorganisms in the environment. There is a unique crosstalk in the microbiome-host interaction. This mutually beneficial symbiotic relationship influences the host physiologic functions (Honda and Littman, 2016). Therefore, regulation of constant gut microbial homeostasis is very important for maintenance of health and requires various mechanistic approaches targeted at separating commensal microorganisms from pathogenic microbes and regulation of intestinal mucosal immunological activities (Honda and Littman, 2016; Levy et al., 2017). This homeostasis can be perturbed by some inflammatory insults or epigenomic manipulations responsible for the dysbiosis that result in impairs of normobiotic microbiota constitution found in gastrointestinal disorders and metabolic diseases (Lepage et al., 2011; Gevers et al., 2014).

The human microbiome composition may now be fully characterized due to the recent advances in the microbiome research. The Human Microbiome Project (NIH HMP, 2009) makes use of sequencing methods to determine the genetic makeup of different gut microbiota. Hence, the entire intestinal microbes can now be elucidated from species to the phyla. Furthermore, the alpha diversity or mean species distribution present in a community can be determined in various ways such as rarefaction curves and species richness indices (Jost, 2007). Also, beta diversity meaning species difference between two varying microbial communities has been employed to show the variation of species in different communities of microbes. Therefore, the microbiota distribution and features have been connected to health status of host. Likewise, disturbed or distorted microbiome (dysbiosis) has been linked to various disease conditions like CDI, metabolic diseases, and fatty liver (van Nood et al., 2013; Lee et al., 2016; Youngster et al., 2016). New therapeutic approaches have emerged for the correction of impaired microbial homeostasis and they include prebiotics, probiotics, and fecal microbiota transplantation (FMT) which has the most potential and efficacy.

FMT involves the transfer of intestinal microbes of a healthy individual into the gastrointestinal tract of a diseased individual so as to change the intestinal microbial ecology and amend the distorted gut homeostasis (dysbiosis) for the treatment of the disease. FMT regularizes the intestinal microbiome community and functions. FMT is, therefore, considered a favorable beneficial procedure for management of ulcerative colitis patients, due largely to the reinstatement of gut microbiome homeostasis (normobiosis) (Khoruts et al., 2010; van Nood et al., 2013; Seekatz et al., 2014; Weingarden et al., 2015). Hence, FMT can be best described as a low-cost, efficient, and harmless technique for therapeutic supplementation of human colonic microbiota for treatment of gastrointestinal disorders. FMT as a concept was further buttressed by a study that revealed the ability of a dysbiotic gut microflora or single commensal bacteria to cause gut inflammation when transported in a germ-free rodent (Ivanov et al., 2009).

The route of administration or infusion of FMT includes the following:

1. Oral pill (Youngster et al., 2014, 2016)
2. Colonoscopic method (Allegretti et al., 2014)

3. Nasogastric or gastrostomy tube (Agilli and Ilga, 2015; Kronman et al., 2015)
4. Post-pyloric method (Wonderlick and D'Agostino, 2016)
5. Retention enema (Eiseman et al., 1958)
6. Upper tract endoscopy (EGD)

A recent study revealed that duodenal injection of feces of a healthy donor into a diseased is as efficient administering treatment via the colon for the treatment of *Clostridium difficile* infection (CDI). However, very little reports exist on the comparison of routes of infusion for many reasons. The most preferred route is possibly dependent on the anatomical locality of the disease. Duodenal infusion as an FMT mode of delivering the donor's fecal material has been reported to be harmless and effective for management of metabolic disease conditions (Vrieze et al., 2012).

Traditionally, antibiotics were used for the management of CDI but they are associated with disruption of intestinal microbiota population which routinely inhibit enlargement and increase of *Clostridium difficile. Clostridium difficile* spores resilient to traditional antibiotics. FMT is a substitute management to usual antibiotic treatment for recurrent CDI for reinstating the intestinal microbiome community of recurrent CDI patients (Brandt and Aroniadis, 2013; De Leon et al., 2013).

5.2 FMT COMPLICATIONS

Safety nature of FMT has been described by numerous studies, for instance, over 200 FMTs have been carried out successfully in the Amsterdam Medical Center, whereas 3,000 FMTs were also done in Australia. These respective procedures were devoid of any severe side effects (Borody and Khoruts, 2012; Vrieze et al., 2012; van Nood et al., 2013). Infusion of fecal materials into most patients result in diarrhea on the same day of administration, and very little number of people reported side effects like cramps, etc. (Gough et al., 2011; Kassam et al., 2013). Actually, complications were documented for just 3 out of 317 patients that were treated with FMT. These complications include peritonitis, upper GIT bleeding, and enteritis.

Nasoduodenal route of FMT was reported to have some side effects in Crohn's disease patient and it includes abdominal tenderness and fever, but these adverse effects vanished in all patients after the second day (Vermeire et al., 2016). Furthermore, colonoscopic route of fecal material infusion did not show any adverse events in the patients (Borody et al., 2008). Interestingly, studies have revealed that patients that received FMT do not have any harmful complications (Brandt and Aroniadis, 2012). Regardless of the acclaimed safety of FMT, there are no adequate follow-up studies on long-term basis. Hence, the potential correlation or link between microbiota transplantation and inflammation or cancer needs more investigations.

5.3 THERAPEUTIC USES OF FMT

Maximum clinical application of FMT has concentrated on the management of recurrent CDI. Afterwards, FMT has been employed in the management of individual with inflammatory bowel diseases (IBD) due to CDI. Furthermore, FMT was applied for the management of individuals with only IBD, chronic fatigue disorders, autoimmune disease, cardiovascular disease, and metabolic syndrome (Borody et al., 1989; Andrews et al., 1992; Borody et al., 2011; Wang et al., 2011; Borody et al., 2012; Vrieze et al., 2012).

5.4 DIET AND GUT MICROBIOTA

Dietary manipulation has been reported to be a significant driver of intestinal microbiome diversity and metabolic functions, making the intestinal microorganisms a connection between nutrition and health (Makki et al., 2018; Valdes et al., 2018). Therefore, nutritional consequences human experience are enormous and cannot be overemphasized because it has capacity to reshape the gut microbiome and its functions in mediating the diet-microbiota-host interactions is very clear from the early stage of life when breast milk from lactating mothers enhance the intestinal microbiome development in infants (Charbonneau et al., 2016), after which the gut flora's bacteria community is enlarged due to the commencement of adult's solid food (Laursen et al., 2017), and finally there is reduction in the gut flora commensal bacteria community as a result of advanced age in elderly likely because of decreased food variety (Claesson et al., 2012).

Diets have the ability to interfere with gut microbiota directly by stimulating or inhibiting some of their developmental and metabolic processes such as growth, and energy generation from particular dietary composition provides a unique benefit to this preferred microorganism of the intestinal microbiota. Therefore, dietary supplementation has the capability to alter the amount, viability, and development of gut microbes (Korem et al., 2015). Diet-mediated reshaping of the gut microbiome activity could cause favorable or harmful health effects which may be as a result of the immunoregulatory functions of the altered microbiome, epigenetic modification in the host or variations in the types of fermentation products or metabolites with the capacity to induce localized effects in the intestine or system wide or interorgan effects. Therefore, the diet-microbiota-host interaction health consequences may not mandatorily need modification of the entire microbial composition but specific microbes in the gut may be altered by the dietary manipulations (Smits et al., 2013).

Diet has emerged as very important modulators of the gut microbiome activity. Food constituents like carbohydrate, protein and fat, probiotics, prebiotics, food additives, and polyphenols are capable of reshaping the gut microbiome constitution and activity (Wang et al., 2017). Diets rich in fiber have been demonstrated to be beneficial in ameliorating metabolic diseases, gastrointestinal disorders, and asthma (Koh et al., 2016). Consumption of various dietary nutrients influences the gut microbiota function, constituents, and host health through diverse mechanistic approaches.

Microbial fermentation of dietary fibers caused formation of metabolic end products including hydrogen, carbon dioxide, and short chain fatty acids (SCFAs) which have enormous roles in reshaping the gut ecology (Chassard and Lacroix, 2013). The intake of fiber-rich plants has been associated with an improved microbial community with abundant *Roseburia, Lachnospira, and Prevotella* with elevated SCFAs generation (De Filippis et al., 2015). However, report in obese humans on low dietary fibers resulted in altered intestinal microbial flora with reduced quantity of gut microbes such as *Roseburia spp., Eubacterium rectale, and Bifidobacteria* alongside proportionate reduction in SCFAs (butyrate) levels in the feces of the subjects (Duncan et al., 2007). Therefore, dietary fiber-mediated alteration in gut microbiome constitutes and increased metabolic end products have the potential to autonomously control the end results of diseases.

Interestingly, dietary fibers have been shown to positively modify or reshape intestinal microbial flora across generations as demonstrated by improved variety and richness of commensal microbes in the intestinal of overweight pregnant women following consumption of fibers (Roytio et al., 2017). Sonnenburg et al. (2016) revealed that consuming diets low in fibers resulted in persistence of negatively altered intestinal microbial community across multiple generations of rodents. The fascinating thing here is that the observed effect cannot be reversed in the offspring even with improved consumption of fiber-rich diets. This clearly emphasizes the developmental origin of diseases in adulthood due to unfavorable conditions during pregnancy on the fetus.

High-fat diet (HFD) supplementation has been associated with the pathobiology of cardiometabolic disorders and obesity. HFD-associated diseases have been connected to alteration of the intestinal microbiome. Butyrate a fermentation product was reported to reverse cardiac damage induced by HFD through uric acid dependent mechanistic pathway (Badejogbin et al., 2019). Fat-rich diets capable of reshaping gut microbiome negatively (Collins et al., 2016) result in inflammation. Gut microbiota production of SCFAs like butyrate was reported to be lowered by HFD probably due to the ability of the HFD to alter the intestinal microbial flora (Cheng et al., 2016).

FMT was reported to alleviate HFD-induced fatty liver through the enhanced modulation gut microbial flora (Zhou et al., 2017). HFD can stimulate the development of a pro-inflammatory gut microbiota which results in elevated permeability of the intestine causing increased plasma levels of lipopolysaccharides (LPS) (Bibbò et al., 2016). Long-term consumption of HFD has also been implicated in the development of neurodegenerative diseases which may be through alteration of the gut microbiota. This interaction has been termed microbiota-gut-brain interaction (Proctor et al., 2016).

Obesity induced by HFD has been reported to be alleviated by sodium butyrate an SCFA through the reversal of gut dysbiosis, decreased circulating LPS levels and reshaping of the intestinal microbiota (Fang et al., 2019). Therefore, HFD is able to modulate intestinal microbiota community and functionality resulting in development of chronic diseases like obesity, cognitive decline, cardiac damage and hepatic steatosis, and even fatty pancreas. This further underscores the potential of diet to manipulate the intestinal microbiota to initiate disease development.

5.5 DONOR SCREENING FOR FMT

The recent status of FMT donor choice basically concentrates on screening for infections together with using a single donor per person method. Questionnaire was used to access risk of infection potency in the donor (Bakken et al., 2011). The FMT research platform OpenBiome and stool banking has advanced screening of infection screening through the use of persistent infection test, approved donors are used continuously, and availability of ready-made resources for transplantation, thereby avoiding the single donor per person method. Concurrently, microbiota solutions are being packaged and authenticated for marketable purposes, especially for persistent CDI, as a ready-made preparation, which disables the challenges associated with fresh FMT infusion (Orenstein et al., 2016).

The foremost purpose of FMT is restoration of normobiosis with proliferation of commensal intestinal microorganisms because of the connection of the microbiome with host physiology and pathophysiology. However, there is need for optimization of the FMT donation procedures due to the fact that not very many donors (less than 10%) can scale through the screening for infection and dysbiosis risk (Paramsothy et al., 2015). Therefore, well-standardized and approved preselected donors may be beneficial considering the challenges. Hence stool samples collection or gathering from numerous donors has been considered for production of transplant materials (O'Donnell et al., 2016). The application of stool from donor is employed for the treatment of CDI. Autologous stool transplant showed increased steadfastness and ease when compared with FMT donation (Kelly et al., 2016).

Another area of importance for research is the implication of donor's diets on the quality, abundance/constitution, and activities of intestinal microbes considering the fact that diets can reshape the gut microbiome which either influence the host positively or negatively (Sonnenburg and Bäckhed, 2016). Therefore, when considering the makeup of vigorous gut microbiome, some factors influence the gut health such as enhanced microbiota diversity and proliferation or abundance of microbes with butyrate-producing potentials (Estaki et al., 2016). The present status of FMT donor selection is still very young, with a lot of questions begging for answers. The future focus must be on the donor with the intention of providing a detailed direction on the enhancement of intestinal microbial community.

Some other vital considerations for donor selection are as follows:

Microbiota diversity indices have been shown to be very critical even after screening for infection risks and particular phyla. Elucidation of typical gut microbiota diversity can give room for detection of abnormality or any aberration in the normal gut microflora via the use of microbiota diversity indices (Halpin and McDonald, 2016).

Gut microbiota nonbacterial members screening may offer more information into the health of the donors. The application of microbiome transplant for CDI clinical management is associated with recent cutting-edge microbiome research through DNA sequencing techniques, genomic and epigenomic approaches to decipher the intestinal microbiota homeostasis or dysbiosis especially for CDI treatment. But there are no sufficient information on the functions of the microbial family excluding bacteria which includes parasites, archaea, fungi, and viruses. These microbes have influences on the host physiology but they have not been completely elucidated. Interestingly majority of reports or investigations concerning evaluating the viral constituent in the microbiome revealed that phages constitute the great number of the viruses in both healthy and dysbiotic persons (Reyes et al., 2010; Norman et al., 2015).

Furthermore, screening for obesity and cardiometabolic syndrome may provide more information alongside the test for pathogenic microbes. Many FMT donation centers reject donors that have obesity, cardiometabolic disorders, or malnourished individuals. Alang and Kelly (2015) reported on the importance of FMT donor selection to the development of obesity during treatment of recurrent CDI who became obese due to donation of fecal materials from an obese but healthy donor. Although donors are now tested to determine if they are malnourished and have cardiometabolic dysfunction resulting in compulsory rejection of donors having cardiometabolic dysfunction also due to the connection of the gut microbes with chronic metabolic diseases (Senthong et al., 2016).

5.6 FECAL MICROBIOTA DELIVERY METHODS

Meighani et al. (2016) reported that the safety and effective delivery of FMT was done through percutaneous endoscopic gastrostomy tube in upper GI. The group also observed that FMT administration in the upper GI may be specific to some group of patient that have contraindications, complication in tolerating lower GI endoscopy, and retention of enemas may not be possible (Meighani et al., 2016). In their study, a smaller proportion of the suspension of fecal was administered to the lower and upper GI tract with volume from 25 to 150/250–500mL for comparison in the level effectiveness and safety. The authors also described that serious aspiration pneumonitis is likely disorders that could results from high volume of 500mL FMT administered through nasoduodenal tube.

In recent time, the therapy used in the prevention of persistent CDI is fecal microbiome transplantation (Mullish et al., 2018) which is an appropriate technique that allows safe prevalent of CDI such as the provision of preprocessed and screened resources locally stored and referred to as stool banking (Smith et al., 2014). The application of this stool banking makes fresh stool to have same clinical efficacy in CDI with the fresh stool (Lee et al., 2016).

However, Allegretti and colleagues in 2020 documented a report that speeds in processing and storage time didn't affect effectiveness of fecal microbiome transplantation. The authors concluded that it was the first large-scale experiment on stool processing and storage time on the use of FMT in the management of CDI. It was also claimed that their finding is in line with an in vitro experiment that support that processing and storage variant of fecal microbiome transplantation support its usage as a guide for the recent best protocol for stool banking (Allegretti et al., 2020). Cammarota et al. (2019) suggest that 6 hours could be recommended as an appropriate stool processing time in fecal microbiome transplantation.

Consequently, fecal microbiome transplantation could be of great benefit in the development of microbiome therapy. However, the complex pathogenesis resulting from bowel disorders from FMT is yet to be elucidated with replacement with therapeutic microbiome. This, therefore, calls for stool banking for FMT (van Lingen et al., 2019). van Lingen and colleagues conducted experiment that dealt with the effectiveness of FMT for management of other dysbiosis linked disorders. The preparation of feces suspension of screened healthy donors was carried out in the stool bank and was quarantined at −80°C after which standardization occurs for quality improvement comparison (van Lingen et al., 2019).

Also, the FMT is gradually becoming an option of novel therapeutic in disorders in which microbiota changes have been involved in disease development and progression (Ooijevaar et al., 2019). Reports have shown that fecal microbiome transplantation is a potent treatment in CDI when used through nasogastric tube (Figure 5.1) (Hagel et al., 2016), nasoduodenal tube (Cohen et al., 2016), enteroscopy (Hagel et al., 2016), and gastroscope-induced infusion (Rubin et al., 2013). Moreover, Hvas et al. (2019) demonstrated superiority of FMT to Fidaxomicin or vancomysin therapy for patient with intractable CDI. Mullish et al. (2018) conducted a randomized experiment on the comparison of FMT and vancomycin in nasoduodenal donor and their report shows the effectiveness of fecal microbiota transplantation over vancomycin (van Nood et al., 2013).

Moreover, Lee et al. (2016) identified intractable CDI to be one of the main concerns to human health in most community due to some associated clinical conditions. However, FMT is recognized for its availability in prompt intervention. A study was then conducted to compare a frozen and fresh fecal microbiome transplantation and the observed results prove that frozen FMT positive for resolution of diarrhea (Lee et al., 2016).

5.7 APPLICATION OF FMT

The FMT is used for restoration of normobiosis in patients with dysbiosis especially those with gastrointestinal disorders, so as to cause maintenance of homeostasis of the microbial flora in the intestine because of the role it plays in physiologic functions of the host (Smits et al., 2013).

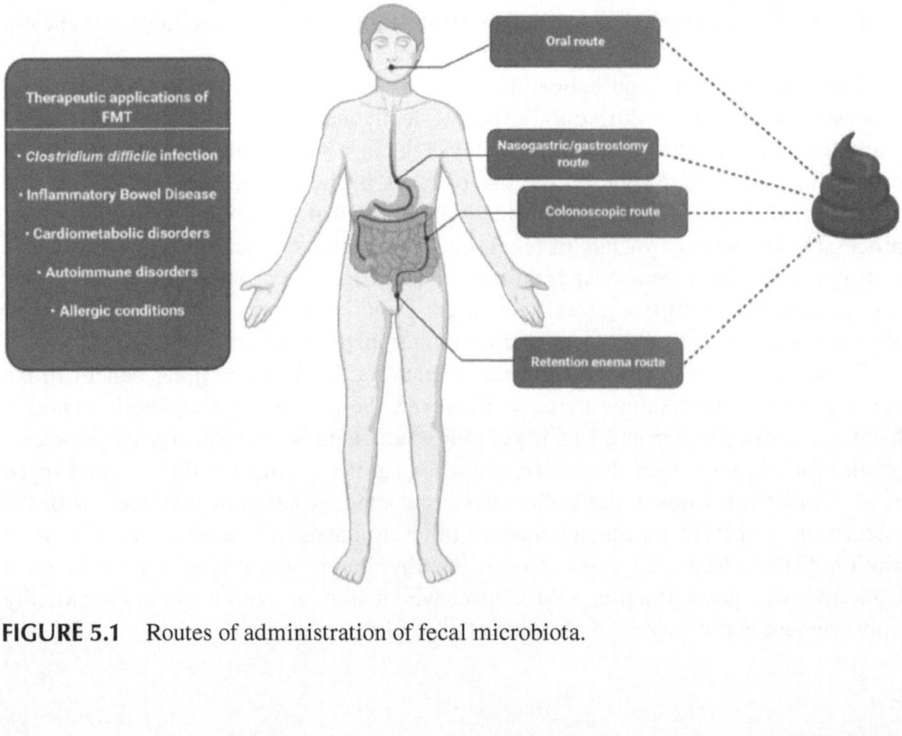

FIGURE 5.1 Routes of administration of fecal microbiota.

5.8 CLOSTRIDIUM DIFFICILE INFECTION

Clostridium difficile is a gram-positive, spore-forming pathogenic bacterium. It is a foremost gastrointestinal infection that has been associated with increased mortality for the last 20 years. This may be due to high spread of hypervirulent strains of Clostridium difficile and inappropriate use of antibiotics (Miller et al., 2011). CDI is a common cause of diarrhea requiring the use antibiotics, and its pathology is facilitated by toxigenic molecules produced by bacteria (Abt et al., 2016). Increasing evidence has established that FMT is efficient for the management of intermittent CDI (van Nood et al., 2013; Kassam et al., 2013; Drekonja et al., 2015; Lee et al., 2016). FMT is a preferred alternative therapy for CDI (Figure 5.1). Hvas and colleagues in their study demonstrated the efficacy and superiority of FMT over traditional antibiotics in the management of intractable CDI (Hvas et al., 2019).

5.9 INFLAMMATORY BOWEL DISEASE

IBD is a disorder of the gut that is associated with protracted inflammation of the gut such as Crohn's disease and ulcerative colitis. IBD has a repetitive attribute of infection advancement and decrease (Gupta et al., 2016). IBD is postulated to take place because of incessant unsuitable antigen activation of gut mucosa lymphatic system (Zhang and Li, 2014). Anderson et al. (2012) reported that FMT therapy for IBD patients showed that 63% of ulcerative colitis patients were relieved, 76% could stop taking drugs for IBD, and 76% had reduced gut inflammations and other associated signs (Figure 5.1).

5.10 CARDIOMETABOLIC DISORDERS

Microbiome transplantation for metabolic dysfunction is such as hypertension, diabetes, and obesity are very important since gut dysbiosis has been associated with the pathogenesis of most metabolic dysfunctions. Microbiome analysis has revealed a close proximity between the microbial community of frank hypertensive and prehypertensive subjects with observed reduced microbiome diversity (Li et al., 2017). Kim and colleagues reported elevated blood pressure in germ-free rodents that had fecal materials infused into them from hypertensive mice, suggesting the significant impact of altered microbiome community in the pathobiology of hypertension (Kim et al., 2018). FMT has been reported to improve insulin sensitivity and glucometabolic regulations in human subjects with metabolic dysfunction that received fecal material from health lean donors (Figure 5.1). This improvement was associated with enhanced bacterial diversity and improved microbiome quality (Vrieze et al., 2012).

5.11 VAGINAL MICROBIOTA TRANSPLANTATION

The great achievement recorded so far on the FMT for the management of CDI has opened up the significance for microbiome transplant as a therapeutic option for numerous diseases with their various conditions. In view of the aforementioned, Delong et al. (2020) identified that vaginal microbiota transplantation (VMT) is the "the next frontier" and also pointed out process involved in bioethics, regulatory, medical, most especially to this nascent field. They also forestall what they considered that will be principal framework in evaluating VMT most especially for clinical investigations, mitigation of relapse of vagina infection. Comparison was made by the authors between vaginal microbiota transplantation with FMT and consequently a statement made on the similarity or variability of the FMT and VMT as regards decision and policy making.

Bacterial vaginosis (BV) has been established to be responsible for vaginal dysbacteriosis together with numerous developments of pathogenic bacteria which could be referred to as a pathological condition of the vagina. Moreover, its management utilizing the antibiotics such as Clindamycin and metronidazole has been discovered to be responsible for high level of reappearance rates. Also, taking into consideration of similar physiological environ around the vaginal tract and intestinal track together as the pathological modes of action they employed during infection of the virginal and the intestine. Therefore, Ma et al. (2019) proposed that the formation of VMT and mention the significant of BV. The authors paid more attention on the pathobiology of BV and the possible adverse effect that it could cause through standardized treatment. They also stated that diverse gut microbiota and tremendously dynamic constitute the most intensive microbial system and also form the most momentous function in human body and it was established that lactobacilli portends that potential to possess the vaginal tract most especially in women, by maintaining their health status. Furthermore, the authors also proposed the methodology of VMT that is based on the influence of FMT most especially toward the management of the intestine infections and list the significant challenges that could hamper the execution of the process involved in the implementation of VMT.

Lev-Sagie and coworkers made pronouncement about the first investigation that validates the application of VMT most especially those from hale and hearty donors who could serve as a beneficial substitute most especially for individuals affected by recurrent BV, intractable and symptomatic. The authors discovered that one of the patients has an incomplete remission most especially during the laboratory and the clinical features while there were no observable influences in all the five women. Moreover, it was observed that remarkably most especially three patients imposed repetitive VMT, which entails a single person making donation to one patient, which provokes an enduring clinical rejoinder. Effectiveness of VMT relapsing BV requires more investigation (Lev-Sagie et al., 2019).

Vaginosis could be identified through the alteration that occurs in the vaginal microbiome and the healing alternatives restricted most especially when it comes to recurrent BV. Some scientists have validated the tendency of the VMT derived from hale and hearty donors for the management of individuals affected from protracted remission and single case which involve partial resolution following VMT. VMT is therefore pointed out as a probable favorable means of managing the disorder. The authors investigated some samples obtained from VMT recipients and donors utilizing shotgun metagenomic sequencing which indicated that four out of five recipients evaluated demonstrated substantial modification in the configuration of the microbiota.

Vaginal restoration is necessary for women with congenital absolute uterine factor infertility (AUFI) and it has been identified as effective optimization of their sexual role. It has been identified that most of these patients do not possess procreative potential, while minimum deliberation has been dedicated to the subsequent vaginal microbiota. Jones et al. (2020) demonstrated that uterine transplantation (UT x) is a promising prospect for provision of effective vaginal restoration. The process involved in the restoration of vagina influence vaginal microbiota composition, hence it is necessary to assess vaginal microbiota in subsequent UT x to comprehend its interrelation with the vaginal microbiota as well as its reproductive and clinical implications.

The great feats recorded during the FMT in effective management of intractable CDI has led to increasing anticipation concerning the use of human materials in transplantation for the treatment of different types of diseases as well as their conditions which are associated to microbial dysbiosis. The level of microbiota available in the vagina in "optimal" condition could identify simply as the vaginal microbiota community, most especially those that are subjugated by very small species of Lactobacillus. It has been shown that the various microbial diversities as well as the associated removal of lactobacilli could be called bacterial vaginosis (BV) that has a relationship with different types of poisonous effects, which entails enhanced possibility of getting sexually transmitted diseases and enhanced possibility of a premature delivery. The authors stated that there are limited small management alternative that could be utilized for their management most especially for the "resetting" the vaginal microbiome to a further beneficial status. Therefore, to the effectiveness of the hypothesis that VMT is a better therapeutic alternative, the authors evaluate the procedure involved in the screening of donors and find out how those subjected to screening could possess little danger of pathogenic microbes transference and maximum vaginal microbiome especially for transplantation purposes. The authors proposed a universal donor screening technique that could be utilized most especially in a small trial with 20 women. The authors later identified major key structural and biochemical features of donor vaginal secretions and the configuration of the corresponding vaginal microbiome in order to demarcate exclusion and inclusion criteria. In addition, the authors also proposed a framework that could help in rapid clinical investigation of VMT.

5.12 INTESTINAL MICROBIOTA TRANSPLANTATION

Garza-González et al. (2019) performed a comparative investigation on how to utilize the variations in the healing effectiveness and intestinal microbiome of FMT using the *Lactobacillus* (FMT-L) for the management of intractable CDI. The authors also performed a 90 days monitoring of 21 patients by collecting their fecal samples on the day 0 to 3rd, 7th, and 28th days after applying treatment. The authors also performed FMT, and fecal samples were evaluated using 16S rRNA sequencing. The results obtained indicated that 13 patients belong to the FMT and 8 to FMT-L out of the 21 patients tested. The general outcome showed that both groups exhibited decrease in the level of the bowel movements per day. Moreover, there were no recurrences and severe contrary rejoinders that were recorded. The presence of *Firmicutes* was documented to have the highest phylum in donors while minimum number of *Proteobacteria* was found in the patients even at greater proportions when compared to the donor. It was discovered that the donor pool entails small *Bacteroidetes*, while some individuals exhibited greater richness. Moreover, results obtained from the

ANOSIM R values indicated momentous modification between the microbial communities most especially from the basal samples as well as samples obtained on the day 7 and at day 28 (0.041). Their study indicated that FMT most especially using the capsules was genomically and clinically alike among improved FMT with *Lactobacillus spp.* and traditional FMT. Also, the reinstatement of microbial diversity and the tenacity of impaired microbiome homeostasis were reported. It was reported that patients having first incident of reappearance treated with FMT demonstrated excellent reaction without having any severe adverse events.

Wang et al. (2014) observed that intestine microbiota performs essential function most especially in the gastrointestinal disorders. Their study showed that IBD pathobiology entails incongruous constant stimulation of the immunological system present in the gut mucosal motivated by altered intestinal microbiota most especially in persons with genetic predisposition. Moreover, there are no precise pathogenic microbes that could be associated with IBD pathogenesis. Intestinal microbial composition together with their metabolic products undeniably distressed in persons afflicted by IBD. The special changes of the gut microbiota related with the IBD need to be investigated. Active IBD predisposes to *Clostridium difficile* due to their interconnected microbial relationship. Moreover, the FMT associated barriers, entailing the modes of reestablishing gut microbiota, preparation of the fecal material, standardized donor screening, and administration, as well as enduring well-being must be determined. Intestinal microbiome function as well as the FMT in the IBD has to be inspected further using metatranscriptomic and metagenomic analysis.

Aroniadis et al. (2014) provided more insights on human microbiota by revealing the impact of science most especially surveillance, improvement of disease treatment as well as how human health function daily could be linked to the microorganism dwelling in intimate associated cell in the body.

The intestinal microbiota has been observed to play a crucial function toward effective regulation of host resistance. The comparative wealth of the microbiota has been observed to depend on environmental and generic factors. Some examples of microorganism such as enterohemorrhagic *E. coli*, *Citrobacter rodentium*, and enteropathogenic *Escherichia coli* have been discovered to be responsible for diarrheal disease. Wang et al. (2018) discovered that microbiota moves could alter both the advantages of manifesting expression of NleH to *C. rodentium* and changes in the level of resistance to the host to the *rodentium* infection most especially in the intestinal colonization.

5.13 LIMITATIONS OF THE FMT

Dangers of transferring suboptimal clones that escaped detection at the metagenomic screening level exist during FMT resulting in pathogenic infection, death, autoimmune diseases, etc. Less severe side effects of FMT include fever, pains and/or cramps in the abdomen, vomiting, constipation, flatulence, belching, and elevated stool frequency (Wang et al., 2016).

Unintended traits could be transferred in the process of treating a given disorder. Studies with genetically similar mice revealed that FMT from an obese mouse was able to treat intestinal dysbiosis of the recipient but made the otherwise lean mouse to become obese (Rodriguez et al., 2019). The limited understanding of the mechanistic role of each operational unit or taxon with the human gut grossly undermines the scientific application of the technology.

5.14 FUTURE TRENDS OF MICROBIOTA TRANSPLANTATION

Employing metagenomics and functional genomics, it is possible to begin to decipher the complex web of interactions between the members of the human gut microbiome. It is known that microbial communities contribute 1000X more genes than the host toward the overall well-being of the host. A clearer understanding of the relative roles of these genomes and the factors that influence their relative abundance and effective interactions will become useful in fine-tuning the therapy. Moreover, the need to co-culture and/or isolate these critical microbionts of humanity will allow future medical applications to use the precise consortium or combination of species in treating specific illnesses. It is anticipated that symbiotics which integrates prebiotics and experimentally determined communities of probionts would dominate all facets of medicine in the near future. Literarily all aspect of human health would have defined consortia of microbial prescription preferred over chemical drugs. Fecal transplants will be replaced by specific psychobionts to treat depression and other mental disorders for example. As a result, there is an ever-growing emphasis in re-inventing microbial cell culture and preserving culture lines. Multinational studies have shown that the gut microbiota in rural Nigeria and Africa are some of the most diverse, containing up to 1,000 new highly treasured bacteria responsible for nutritional, mental, and overall healthy humans.

5.15 CONCLUSION

The unique influence of gut microbiome in maintaining host health cannot be overemphasized because gut dysbiosis is connected to the pathogenesis of some chronic diseases. Interesting FMT therapy is a hopeful alternative for ameliorating diseases especially some antibiotic-resistant infections like CDI. The normalization of the intestinal microbiota ecology by FMT along with resolution of inflammatory tendencies is essentially responsible for the treatment of gastrointestinal disorders associated with impaired microbiota homeostasis (Figure 5.2). Stool banking also provides ease of storage and timely accessibility for the patients in dire need of FMT. Nigeria and developing countries should invest in creating cell culture collection centers to power the discovery and application of the newest human organ gut microbiome.

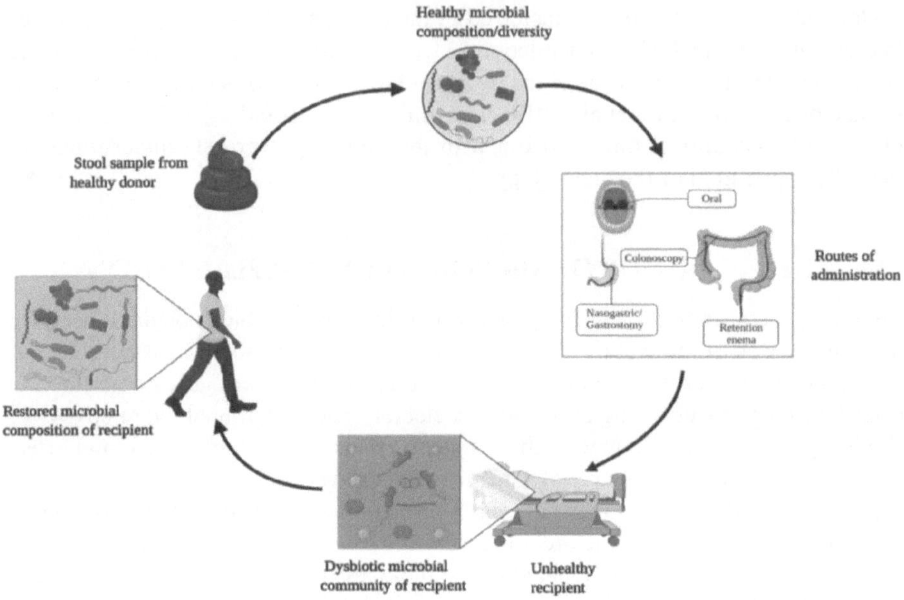

FIGURE 5.2 Schematic illustration of fecal microbiota transplantation from healthy donor to unhealthy recipient.

REFERENCES

Abt M.C., McKenney P.T., Pamer E.G. (2016) Clostridium difficile colitis: pathogenesis and host defence. *Nature Reviews Microbiology*; 14(10): 609–620.

Agilli M., Ilga U. (2015) A methodological approach to fecal microbial transplant via nasogastric tube for active pediatric ulcerative colitis. Journal of Pediatric Gastroenterology and Nutrition; 60: e36.

Alang N., Kelly C.R. (2015) Weight gain after fecal microbiota transplantation. *Open Forum Infectious Diseases* [Internet]; 2: ofv004. Available from: http://ofid. oxfordjournals.org/content/early/2014/06/01/ofid. ofu038.short; PMID:26034755. doi: 10.1093/ofid/ofv004.

Allegretti J.R. et al. (2020) Stool processing speed and storage duration do not impact the clinical effectiveness of fecal microbiota transplantation. *Gut Microbes*. doi: 10.1080/19490976.2020.1768777.

Allegretti J.R., Korzenik J.R., Hamilton M.J. (2014) Fecal microbiota transplantation via colonoscopy for recurrent C. difficile infection. *Journal of Visualized Experiments (JoVE)*. doi: 10.3791/52154.

Anderson J.L., Edney R.J., Whelan K. (2012) Systematic review: faecal microbiota transplantation in the management of inflammatory bowel disease. *Alimentary Pharmacology & Therapeutics*; 36(6): 503–516.

Andrews P., Barnes P., Borody T. (1992) Chronic constipation reversed by restoration of 17 bowel flora: a case and a hypothesis. *European Journal of Gastroenterology and Hepatology*; 4: 245–247.

Aroniadis O.C., Brandt L.G., AGA-F. (2014) Intestinal microbiota and the efficacy of fecal microbiota transplantation in gastrointestinal disease. *Gastroenterology and Hepatology*; 10(4): 230-237.

Badejogbin C., Areola D.E., Olaniyi K.S., Adeyanju O.A., Adeosun I.O. (2019) Sodium butyrate recovers high-fat diet-fed female Wistar rats from glucose dysmetabolism and uric acid-associated cardiac tissue damage. *Naunyn-Schmiedeberg's Archives of Pharmacology*; 392(11): 1411–1419.

Bakken J.S. et al. (2011) Treating Clostridium difficile infection with fecal microbiota transplantation. *Clinical Gastroenterology and Hepatology*; 9: 1044–1049.

Bibbò S., Ianiro G., Giorgio V., Scaldaferri F., Masucci L., Gasbarrini A., Cammarota G. (2016). The role of diet on gut microbiota composition. *European Review for Medical and Pharmacological Sciences*; 20: 4742–4749.

Borody T.J. et al. (1989) Bowel-flora alteration: a potential cure for inflammatory bowel 10 disease and irritable bowel syndrome? *The Medical journal of Australia*; 150: 604–611.

Borody T. et al. (2008) Clostridium difficile complicating inflammatory bowel disease: pre and post-treatment findings. *Gastroenterology*; 134: A361.

Borody T.J., Khoruts, A. (2012) Fecal microbiota transplantation and emerging applications. Nature reviews. *Gastroenterology and Hepatology*; 9: 88–96.

Borody T.J., Leis S.M., Campbell J.L., Torres M. (2011) Fecal microbiota transplantation (FMT) in multiple sclerosis (MS). *American Journal Gastroenterology*; 106: 842–843.

Borody T., Nowak A., Finlayson S. (2012) The GI microbiome and its role in chronic fatigue syndrome: a summary of bacteriotherapy. *Australasian College of Nutritional and Environmental Medicine*; 31: 3–8.

Brandt L., Aroniadis, O. (2012) Long-term follow-up study of fecal microbiota transplantation (FMT) for ulcerative colitis (UC). ACG 2012 Annual Scientific Meeting Abstracts P405.

Brandt L.J., Aroniadis O.C. (2013) An overview of fecal microbiota transplantation: techniques, indications, and outcomes. *Gastrointestinal Endoscopy*; 78: 240–249.

Cammarota G., et al. (2019) International consensus conference on stool banking for faecal microbiota transplantation in clinical practice. *Gut*; 68(12): 2111–2121. doi: 10.1136/gutjnl-2019-319548.

Charbonneau, M.R. et al. (2016) Sialylated milk oligosaccharides promote microbiota-dependent growth in models of infant under nutrition. *Cell*; 164, 859–871.

Chassard C., Lacroix C. (2013) Carbohydrates and the human gut microbiota. *Current Opinion in Clinical Nutrition and Metabolic Care*; 16(4), 453–460.

Cheng L. et al. (2016) High fat diet exacerbates dextran sulfate sodium induced colitis through disturbing mucosal dendritic cell homeostasis. *International Immunopharmacology*; 40: 1–10.

Claesson M.J. et al. (2012) Gut microbiota composition correlates with diet and health in the elderly. *Nature*; 488(7410): 178–184.

Cohen N.A., Livovsky D.M., Yaakobovitch S., Ben Yehoyada M., Ben Ami R., Adler A., Guzner-Gur H., Goldin E., Santo M.E., Halpern Z., Paz K., Maharshak N. (2016) A retrospective comparison of fecal microbial transplantation methods for recurrent Clostridium difficile infection. *The Israel Medical Association Journal*; 18(10): 594–599.

Collins K.H., Paul H.A., Hart D.A., Reimer R.A., Smith I.C., Rios J.L., Herzog W. (2016) A high-fat high-sucrose diet rapidly alters muscle integrity, inflammation and gut microbiota in male rats. *Scientific Reports*; 6: 37278.

De Filippis F., Pellegrini N., Vannini L., Jeffery I.B., La Storia A., Laghi L., Ercolini D. (2015) High-level adherence to a Mediterranean diet beneficially impacts the gut microbiota and associated metabolome. *Gut*; 65(11): 1812–1821.

De Leon L.M., Watson J.B., Kelly C.R. (2013) Transient flare of ulcerative colitis after fecal microbiota transplantation for recurrent Clostridium difficile infection. *Clinical Gastroenterology and Hepatology*; 11: 1036–1038.

DeLong K., Zulfiqar F., Hoffmann D.E., Ensign L.M. (2020) Vaginal microbiota transplantation: the next frontier. *Journal of Law, Medicine & Ethics*. doi: 10.1177/1073110519897731.

Drekonja D., et al. (2015) Fecal microbiota transplantation for Clostridium difficile infection a systematic review. *Annals of Internal Medicine*; 162(9): 630–638.

Duncan S.H., Belenguer A., Holtrop G., Johnstone A.M., Flint H.J., Lobley G.E. (2007) Reduced dietary intake of carbohydrates by obese subjects results in decreased concentrations of butyrate and butyrate-producing bacteria in feces. *Applied Environmental Microbiology*; 73(4): 1073–1078.

Eiseman B., Silen W., Bascom G.S., Kauvar A.J. (1958) Fecal enema as an adjunct in the treatment of pseudomembranous enterocolitis. *Surgery*; 44: 854–859.

Estaki M. et al. (2016) Cardiorespiratory fitness as a predictor of intestinal microbial diversity and distinct metagenomic functions. *Microbiome*; 4: 42.

Fang W., Xue H., Chen X., Chen K., Ling W. (2019) Supplementation with sodium butyrate modulates the composition of the gut microbiota and ameliorates high-fat diet-induced obesity in mice. *Journal of Nutrition*; 149(5): 747–754.

Garza-González E., Mendoza-Olazarán S., Morfin-Otero R., Ramírez-Fontes A., Rodríguez-Zulueta P., Flores-Treviño S., Bocanegra-Ibarias P., Maldonado-Garza H., Camacho-Ortiz A. (2019). Intestinal microbiome changes in fecal microbiota transplant (FMT) vs. FMT enriched with Lactobacillus in the treatment of recurrent Clostridioides difficile infection. *Canadian Journal of Gastroenterology and Hepatology*, Article ID 4549298: 1–7.

Gevers D. et al. (2014) The treatment-naive microbiome in new-onset Crohn's disease. *Cell Host Microbe*; 15: 382–392.

Gough E., Shaikh H., Manges A.R. (2011) Systematic review of intestinal microbiota transplantation (fecal bacteriotherapy) for recurrent Clostridium difficile infection. *Clinical Infectious Diseases: An Official Publication of the Infectious Diseases Society of America*; 53: 994–1002.

Gupta S., Emma Allen-Vercoe and Elaine O. (2016) Petrof Fecal microbiota transplantation: in perspective. *Theraupetic Advances in Gastroenterology*, 9(2): 229–239.

Hagel S., Fischer A., Ehlermann P., Frank T., Tueffers K., Sturm A., Link A., Demir M., Siebenhaar A., Storr M., Glueck T., Siegel E., Solbach P., Goeser F., Koelbel C.B., Lohse A., Luebbert C., Kandzi U., Maier M., Schuerle S., German Clinical Microbiome Study Group (GCMSG) (2016) Fecal microbiota transplant in patients with recurrent Clostridium Difficile infection. *Deutsches Arzteblatt international*; 113(35–36): 583–589. doi: 10.3238/arztebl.2016.0583.

Halpin A.L., McDonald L.C. (2016) Editorial Commentary: The dawning of microbiome remediation for addressing antibiotic resistance. *Clinical Infectious Diseases* [Internet]; 62: 1487–1488. Available from: http://cid.oxfordjournals.org/ lookup/doi/10.1093/cid/ciw187; PMID: 27025827. doi: 10.1093/cid/ciw187.

Honda K., Littman D.R. (2016) The microbiota in adaptive immune homeostasis and disease. *Nature Cell Biology*; 535: 75–84.

Hvas C.L., Dahl Jørgensen S.M., Jørgensen S.P., Storgaard M., Lemming L., Hansen M.M., Erikstrup C., Dahlerup J.F. (2019). Fecal microbiota transplantation is superior to fidaxomicin for treatment of recurrent Clostridium dufficile infection. *Gastroenterology*, 156(5):1324–1332.e3.

Ivanov I.I., Atarashi K., Manel N., Brodie E.L., Shima T., Karaoz U., Wei D., Goldfarb K.C., Santee C.A., Lynch S.V., Tanoue T., Imaoka A., Itoh K., Takeda K., Umesaki Y., Honda K., Littman D.R. (2009) Induction of intestinal Th17 cells by segmented filamentous bacteria. *Cell*; 139(3): 485–498.

Jones B.P., Saso S., L'Heveder A., Bracewell-Milnes T., Thum M.Y., Diaz-Garcia C., MacIntyre D.A., Quiroga I., Ghaem-Maghami S., Testa G., Johannesson L., Bennett P.R., Yazbek J., Smith J.R. (2020) The vaginal microbiome in uterine transplantation. *BJOG*; 127(2): 230–238.

Jost L. (2007) Partitioning diversity into independent alpha and beta components. *Ecology*; 88: 2427–2439.

Kassam Z., Lee C.H., Yuan Y., Hunt R.H. (2013) Fecal microbiota transplantation for Clostridium difficile infection: systematic review and meta-analysis. *American Journal of Gastroenterology*, 108(4): 500–508.

Kelly C.R. et al. (2016) Effect of fecal microbiota transplantation on recurrence in multiply recurrent Clostridium difficile infection: a randomized trial. *Annals of Internal Medicine*. doi: 10.7326/M16-0271.

Khoruts A., Dicksved J., Jansson J.K., Sadowsky M.J. (2010) Changes in the composition of the human fecal microbiome after bacteriotherapy for recurrent Clostridium difficile-associated diarrhea. *Journal of Clinical Gastroenterology*; 44: 354–360.

Kim T.T. et al. (2018) Fecal transplant from resveratrol-fed donors improves glycaemia and cardiovascular features of the metabolic syndrome in mice. *American Journal of Physiology - Endocrinology and Metabolism*; 315: E511–E519. doi: 10.1152/ajpendo.00471.2017.

Koh A., De Vadder F., Kovatcheva-Datchary P., Backhed F. (2016) From dietary fiber to host physiology: short-chain fatty acids as key bacterial metabolites. Cell; 165. doi: 10.1016/j.cell.2016.05.041.

Korem, T. et al. (2015) Growth dynamics of gut microbiota in health and disease inferred from single metagenomic samples. *Science*, 349: 1101–1106.

Kronman M.P. et al. (2015) Fecal microbiota transplantation via nasogastric tube for recurrent Clostridium difficile infection in pediatric patients. *Journal of Pediatric Gastroenterology and Nutrition*; 60: 23–26.

Laursen M.F., Bahl M.I., Michaelsen K.F., Licht T.R. (2017) First foods and gut microbes. *Frontiers in Microbiology*; 8: 356.

Lee C.H., Steiner T., Petrof E.O., Smieja M., Roscoe D., Nematallah A., Weese J.S., Collins S., Moayyedi P., Crowther M., Ropeleski M.J., Jayaratne P., Higgins D., Li Y., Rau N.V., Kim P.T. (2016). Frozen vs fresh fecal microbiota transplantation and clinical resolution of diarrhea in patients with recurrent Clostridium difficile infection: a randomized clinical trial. *JAMA*, 315(2): 142–149. doi: 10.1001/jama.2015.18098.

Lepage P. et al. (2011) Twin study indicates loss of interaction between microbiota and mucosa of patients with ulcerative colitis. *Gastroenterology*; 141: 227–236.

Lev-Sagie A., Goldman-Wohl D., Cohen Y., Dori-Bachash M., Leshem A., Mor U., Strahilevitz J., Moses A.E., Shapiro H., Yagel S., Elinav E. (2019) Vaginal microbiome transplantation in women with intractable bacterial vaginosis. *Nature Medicine*; 25: 1500–1504.

Levy M., Kolodziejczyk A.A., Thaiss C.A., Elinav E. (2017) Dysbiosis and the immune system. *Nature Reviews Immunology*; 17: 219–232.

Li J. et al. (2017) Gut microbiota dysbiosis contributes to the development of hypertension. *Microbiome*; 5: 14. doi: 10.1186/s40168-016-0222-x.

Ma D., Chen Y., Chen T. (2019) Vaginal microbiota transplantation for the treatment of bacterial vaginosis: a conceptual analysis. *FEMS Microbiology Letters*; 366(4): fnz025. doi: 10.1093/femsle/fnz025.

Makki K. et al. (2018) The impact of dietary fiber on gut microbiota in host health and disease. *Cell Host Microbe*; 23(6): 705–715.

Meighani A., Hart B.R., Mittal C., Miller N., John A., Ramesh M. (2016) Predictors of fecal transplant failure. *European Journal of Gastroenterology and Hepatology*; 28(7): 826–830. doi: 10.1097/MEG.0000000000000614.

Miller B., Chen L., Sexton D., Anderson D. (2011) Comparison of the burdens of hospital-onset, healthcare facility-associated Clostridium difficile infection and of healthcare-associated infection due to methicillin-resistant Staphylococcus aureus in community hospitals. *Infection Control and Hospital Epidemiology*; 32: 387–390.

Mullish B.H., et al. (2018). The use of faecal microbiota transplant as treatment for recurrent or refractory Clostridium difficile infection and other potential indications: Joint British Society of Gastroenterology (BSG) and Healthcare Infection Society (HIS) guidelines. *Gut*; 67(11): 1920–1941. doi: 10.1136/gutjnl-2018-316818.

NIH HMP Working Group et al. (2009) The NIH Human Microbiome Project. *Genome Research*; 19: 2317–2323.

Norman J.M. et al. (2015) Disease-specific alterations in the enteric virome in inflammatory bowel disease. *Cell* [Internet]; 160: 447–460. Available from: http://dx.doi.org/10.1016/j.cell.2015.01.002; PMID: 25619688. doi: 10.1016/j.cell.2015.01.002.

O'Donnell M.M. et al. (2016) Preparation of a standardized faecal slurry for ex-vivo microbiota studies which reduces inter-individual donor bias. *Journal of Microbiological Methods*; 129: 109–116.

Ooijevaar R.E., Terveer E.M., Verspaget H.W., Kuijper E.J., Keller J.J. (2019) Clinical application and potential of fecal microbiota transplantation. *Annual Review of Medicine*; 70: 335–351.

Orenstein R. et al. (2016) Safety and durability of RBX2660 (microbiota suspension) for recurrent Clostridium difficile infection: results of the PUNCH CD study. *Clinical Infectious Diseases*; 62: 596–602.

Paramsothy S. et al. (2015) Donor recruitment for fecal microbiota transplantation. Inflammatory Bowel Diseases; 21: 1600–1606.

Proctor C., Thiennimitr P., Chattipakorn N., Chattipakorn S.C. (2016) Diet, gut microbiota and cognition. Metabolic Brain Diseases. doi: 10.1007/s11011-016-9917-8.

Reyes A., Haynes M., Hanson N., Angly F.E., Heath A.C., Rohwer F., Gordon J.I. (2010) Viruses in the faecal microbiota of monozygotic twins and their mothers. *Nature* [Internet]; 466: 334–338. Available from: http://dx.doi.org/ 10.1038/nature09199; PMID: 20631792. doi: 10.1038/nature09199.

Rodriguez D.M., Benningho A.D, Aardema N.D.J., Phatak S., Hintze K.J. (2019) Basal diet determined long-term composition of the gut microbiome and mouse phenotype to a greater extent than fecal microbiome transfer from lean or obese human donors. *Nutrients*; 11: 1630. doi: 10.3390/nu11071630.

Roytio H., Mokkala K., Vahlberg T., Laitinen K. (2017) Dietary intake of fat and fibre according to reference values relates to higher gut microbiota richness in overweight pregnant women. *British Journal of Nutrition*; 118(5): 343–352.

Rubin T.A., Gessert C.E., Aas J., Bakken J.S. (2013) Fecal microbiome transplantation for recurrent Clostridium difficile infection: report on a case series. *Anaerobe*; 19: 22–26.

Seekatz A.M. et al. (2014) Recovery of the gut microbiome following fecal microbiota transplantation. *mBio*; 5: e00893–e00814.

Senthong V., Wang Z., Li X.S., Fan Y., Wu Y., Wilson Tang W.H., Hazen S.L. (2016) Intestinal microbiota-generated metabolite trimethylamine- N- oxide and 5- year mortality risk in stable coronary artery disease: the contributory role of intestinal microbiota in a COURAGE-like patient cohort. *Journal of American Heart Association*; 5: e002816. Available from: http:// jaha.ahajournals.org/lookup/doi/10.1161/JAHA.115. 002816; PMID: 27287696. doi: 10.1161/ JAHA.115.002816.

Smith M., Kassam Z., Edelstein C., Burgess J., Alm E. (2014) Open biome remains open to serve the medical community. *Nature Biotechnology*; 32: 867. doi: 10.1038/ nbt.3006.

Smits L., Bouter K., De Vos W., Borody T., Nieuwdorp M. (2013) Therapeutic potential of fecal microbiota transplantation. *Gastroenterology*; 145: 946–953.

Sonnenburg J.L., Bäckhed F. (2016) Diet-microbiota interactions as moderators of human metabolism. *Nature*; 535: 56–64.

Sonnenburg E.D., Smits S.A., Tikhonov M., Higginbottom S.K., Wingreen N.S., Sonnenburg J.L. (2016) Diet-induced extinctions in the gut microbiota compound over generations. Nature; 529(7610): 56–64.

Valdes A.M. et al. (2018) Role of the gut microbiota in nutrition and health. *BMJ*; 361: k2179.

van Lingen E., Terveer E.M., van der Meulen-de Jong A.E., Vendrik K.E.W., Verspaget H.W., Kuijper E.J., Kassam Z., Keller J.J. (2019) Advances in stool banking. *Microiota in Health and Disease*; 1: e182.

van Nood E., Vrieze A., Nieuwdorp M., Fuentes S., Zoetendal E.G., de Vos W.M., Visser C.E., Kuijper E.J., Bartelsman J.F., Tijssen J.G., Speelman P., Dijkgraaf M.G., Keller J.J. (2013) Duodenal infusion of donor feces for recurrent Clostridium difficile. *The New England Journal of Medicine*; 368(5): 407–415. doi: 10.1056/NEJMoa1205037.

Vermeire S., Joossens M., Verbeke K., Wang J., Machiels K., Sabino J., Ferrante M., Van Assche G., Rutgeerts P., Raes J. (2016) Donor species richness determines faecal microbiota transplantation success in inflammatory bowel disease. *Journal of Crohn's and Colitis*, 10(4): 387–394.

Vrieze A. et al. (2012) Transfer of intestinal microbiota from lean donors increases insulin sensitivity in individuals with metabolic syndrome. *Gastroenterology*; 143: 913–916.e7. doi: 10.1053/j.gastro.2012.06.031.

Wang Z. et al. (2011) Gut flora metabolism of phosphatidylcholine promotes cardiovascular disease. *Nature*; 472: 57–63.

Wang S., et al. (2016) Systematic review: adverse events of fecal microbiota transplantation. *PLoS ONE*; 11(8): e0161174. doi: 10.1371/journal.pone.0161174.

Wang G., Feuerbacher L.A., Hardwidge P.R. (2018) Influence of intestinal microbiota transplantation and NleH expression on Citrobacter rodentium colonization of mice. *Pathogens*; 7(2): 35. doi: 10.3390/pathogens7020035.

Wang Z.K., Yang Y.S., Chen Y., Yuan J., Sun G., Peng L.H. (2014) Intestinal microbiota pathogenesis and fecal microbiota transplantation for inflammatory bowel disease. *World Journal of Gastroenterology*; 20(40): 14805–14820.

Wang B., Yao M., Lv L., Ling Z., Li L. (2017) The human microbiota in health and disease. *Engineering*; 3(1): 71–82.

Weingarden A. et al. (2015) Dynamic changes in shortand long-term bacterial composition following fecal microbiota transplantation for recurrent Clostridium difficile infection. *Microbiome*; 3: 10.

Wonderlick J.S., D'Agostino R. (2016) Fecal microbiota transplantation via fluoroscopy-guided nasojejunal catheter placement: indications, technique, and the role of radiology. *Abdomical Radiology (N Y)*; 41: 2020–2025.

Youngster I. et al. (2014) Oral, capsulized, frozen fecal microbiota transplantation for relapsing Clostridium difficile infection. *JAMA*; 312: 1772–1778.

Youngster I. et al. (2016) Oral, frozen fecal microbiota transplant (FMT) capsules for recurrent Clostridium difficile infection. *BMC Medicine*; 14: 134.

Zhang Y. and Li Y. (2014) inflammatory bowel disease: pathogenesis. *World Journal Gastroenterology*; 20: 91–99.

Zhou D., Pan Q., Shen F., Cao H.X., Ding W.J., Chen Y.W., Fan J.G. (2017) Total fecal microbiota transplantation alleviates high-fat diet-induced steatohepatitis in mice via beneficial regulation of gut microbiota. *Scientific Reports*; 7(1): 1529.

6 Microbiome Reshaping and Epigenetic Regulation

Olugbenga Samuel Michael, Olufemi Idowu Oluranti, and Ayomide Michael Oshinjo
Bowen University

Charles Oluwaseun Adetunji
Edo State University Uzairue

Kehinde Samuel Olaniyi
Afe Babalola University

Juliana Bunmi Adetunji
Osun State University

CONTENTS

6.1 Introduction 99
6.2 Microbiome, Development, and Epigenetic Programming 101
6.2.1 Diet-Microbiome-Epigenetics Crosstalk 104
6.3 Microbiome, Immunity, and Epigenetic Modification 105
6.3.1 Microbiota, Epigenetics, and Neurodegeneration 110
6.4 Future Perspectives 111
6.5 Conclusion 112
References 113

6.1 INTRODUCTION

Mammals live in continuous contact or close proximity with microorganisms which eventually colonize the gastrointestinal tracts of the host such that they both undergo growth and evolution together. The human intestinal microbial flora is made up of trillions of microorganisms such as viruses, eukaryotes, archaea, and bacteria (Allin et al., 2015). The microbiome is an unfolding area that has recently stimulated research interest in metabolic health. It has links to other chronic diseases like diabetes, cancer, hypertension, etc. through different mechanistic processes such as epigenetic modifications. Cutting-edge, high-throughput metagenomic and metabolomic techniques have exposed the contributory roles of the microbiome to numerous

DOI: 10.1201/9781003180241-6

physiological or biological processes essential for healthy living (Kelly et al., 2016; Sharma et al., 2019).

The intestinal microbiota is colonized by several microbes such as *Actinobacteria, Firmicutes, Tenericutes, Verrucomicrobia, Bacteroidetes, Lentisphaerae, and Proteobacteria* as the leading phyla (Marlicz et al., 2018). Colonization of newborn by microorganisms is a fundamental developmental process that influences health and disease pathogenesis in later life (Stiemsma and Michels, 2018; Korpela and de Vos, 2018; Korpela et al., 2020). Metagenomics analysis and microbial strain profiling demonstrated that the fecal bacteria of the mother are the principal colonizers of the normal infant gastrointestinal tract. They include *Bacteroides and Bifidobacterium* (Nayfach et al., 2016; Asnicar et al., 2018). Apart from the maternal fecal bacteria, a neonate also acquires its microbiota from breastfeeding and the microbes present in the delivery room (Dominguez-Bello et al., 2010). Hence, dietary influences are essential for reshaping the microbiome during developmental stages of life (Cotillard et al., 2013).

Diet has been recognized to have a greater influence on epigenetic alteration related to diseases and portends the capability to change the process associated with the control of the genetic manifestation patterns most especially via epigenetic mechanisms (Huang et al., 2011; Davis and Ross, 2007). The mode of action associated with epigenetics involves the manifestation of the target genes such as noncoding RNAs and histone modifications. Moreover, the process involved in the microbiome-mediated dietary metabolic activities could generate molecules with capacity to modify epigenetic regulations through processes which include, changing the amount of substances utilized for genetic alteration like methylation, as well as production of other substances that could change enzymes (DNA methyltransferases, histone methyltransferases and histone deacetylases) that regulate epigenetics machinery, e.g., changes in the amino acid sequence of these enzymes. Furthermore, it has been observed that these epigenetic signatures can be steadily transferred throughout cellular divisions resulting in heritable modification in gene expression without alteration in DNA sequence (Han et al., 2019). These are essential mechanisms for the regulation of identity of cells and also deliver a learning system that permits cells to respond and acclimatize to environmental modifications (Fischer, 2020). Several authors have presented that infectious pathogens have epigenetic regulatory mechanisms through which they efficiently take advantage of host epigenome for their own needs resulting in numerous influences on human health, and the genotoxicity influence of microbial metabolites has also been validated (Alvarez-Venegas, 2014; Hullar et al., 2014; Fol et al., 2020; Bierne and Hamon, 2020).

Hullar and Fu (2014) provided detailed evidence on the activities of the microbiota particularly in dietary metabolism. They described general overview of the modes of action involved in the eukaryotic epigenetic, as well as the biologically active metabolites that could be derived from microorganism. Moreover, the authors paid more attention to the numerous microbial metabolic products like short-chain fatty acids (SCFA) and significant diet that could assist in shaping the constituents of the intestinal microbiota. Dietary manipulations have the capability to alter microbial configuration within 1 day of commencement with resolution of microbial alteration after 2 days of termination of the dietary regimen. Microbial-derived metabolic products like SCFAs (acetate butyrate and propionate) exhibit physiological signaling functions with lasting impact on human health. For instance, acetate and butyrate have been demonstrated to inhibit histone deacetylase. They also influence glucose homeostasis and insulin sensitivity. Obesity and insulin resistance have been negatively correlated with high circulating level of SCFAs (Lin et al., 2007). Recently, we have demonstrated that administration of SCFAs can mitigate high-fat or nicotine-induced impaired gut homeostasis and cardiometabolic dysregulation. Acetate was shown to ameliorate nicotine-induced cardiorenal dysmetabolism (Michael et al., 2020), while butyrate arrests pancreato-hepatic uric acid and lipid dysmetabolism (Adeyanju et al., 2020). Therefore, it is essential to emphasize that human health can be deliberately altered through dietary manipulation and the effect can be determined through our microbes or microbial metabolites.

Children who reside in Burkina Faso have distinctly different intestinal microbiome composition when compared to their European counterpart especially those within the same age categories (De Filippo et al., 2010). Microbiota of the Burkina Faso children generated more butyrate together with the presence of *Prevotella* when compared to *Bacteroides*. It was established that children living in Burkina Faso ingested a diet that was higher in fiber as well as nonanimal protein but the European children were fed with diet containing better animal protein and carbohydrates. Moreover, similar results have been documented in developed and non-developed countries (Yatsunenko et al., 2012; Lin et al., 2013). Therefore, a relationship exists between microbiome in human populations and the diet. It has been established that there are other factors that could impact the constituents and role of the microbiome in these diverse habitats (Bendiks and Kopp, 2013). Dietary intervention is a critical modifier of the intestinal microbiome composition resulting in the alteration of the physiological processes. Gut microbiota-derived metabolites like SCFAs are crucial signaling molecules involved in the regulation of physiological activities. Gut dysbiosis may be due to dietary lifestyles; this impaired gut microbial homeostasis has been linked to some common disease conditions like obesity, diabetes, hypertension, etc. (Hylla et al., 1998, Johnson et al., 2006, Duncan et al., 2007; Costabile et al., 2008; Russell et al., 2011; Menni et al., 2017).

Therefore, this section intends to make available contemporary trends regarding microbiome reshaping and epigenetics. Special highlights were provided on the relevant process involved in the microbiome reshaping and epigenetic, as well as the fact that significant metabolites derived from gut microbiome have several health benefits in the management of numerous diseases.

6.2 MICROBIOME, DEVELOPMENT, AND EPIGENETIC PROGRAMMING

Prenatal microbiome is involved in the critical early and late pregnancy development. It was then observed in the analysis of different biogeography microbiota in women that the configuration of microbiome differs from nonpregnant to pregnant women and also varied in pregnancy (Koren et al., 2012; Mesa et al., 2020). Microbiota colonize egg capsule in some insects during mating and this eventually leads to the colonization of the insect before the process of hatching (Davidson and Stahl, 2006).

In humans, mating mostly takes place in an immune-sealed environment which lacks colonization though not completely sterile (Dominguez-Bello et al., 2019). However, some microbiota can enter the uterine cervix during fertilization by sperm (Chiu et al., 2017), thus reaching the egg during implantation or fetal development (Dominguez-Bello et al., 2019).

During normal delivery via the vagina, the infant at birth is colonized by bacteria due to the existing maternal-fetal interactions (Robertson et al., 2019). This microbial colonization causes cascades of associated inflammation in preterm birth leading to a loss in fetal weight. The most common species of microbiota that dominates the vagina in the cause of pregnancy is *Lactobacillus* species (Callahan et al., 2017; Chu et al., 2017).

Robertson et al. (2019) further disclosed that vaginal microbiome with deficient *Lactobacillus* species have four basic subtypes. However, geographic and environmental variations seem to have influence on the vaginal microbiome during pregnancy with *Gemella spp.*, *Prevotella spp.*, and *Corynebacterium spp.* being the most prevalent. Studies have shown that the reduction in infant length-for-age is determined by pregnancy duration (Aagaard et al., 2014; Doyle et al., 2017). Maternal vaginal microbiota could also be involved in the prenatal programming that influences growth (Robertson et al., 2019).

It was also documented that the environments during fetal growth have different microbiota while the gut microbiota increases during pregnancy (Perez et al., 2007). One such is the reduction in birth weight and preterm birth caused by gut dysbiosis in inflammatory bowel diseases in maternal pregnancy (Getahun et al., 2014). In addition, Dahl and coworkers (2017) reported that analysis of gut microbiome of mothers with preterm birth is associated with low gut microbial diversity characterized by reduced *Clostridium spp.*, *Bifidobacterium spp.*, and *Streptococcus spp.* It was also revealed that high level of *Actinomyces naeslundii* and *Lactobacillus casei* in the saliva of pregnant women can have both negative and positive impact on the weight of the newborn infant (Dasanayake et al., 2005). Dysbiotic oral microbiome, as seen in periodontal disease, results in adverse pregnancy outcomes, atherosclerosis, etc. (Cobb et al., 2017). There is an evidence of oral-placental microbiome axis in preterm birth (Fardini et al., 2010). Dysbiotic oral microbiome can be translocated to the placenta resulting in preterm birth. Gut and oral microbiomes have contributory influence on the development of placental microbiome (Gomez-Arango et al., 2017). Periodontal disease can induce deadly inflammation to the womb, leading to small-for-gestational age (SGA) or premature delivery through microbiota-dependent mechanism.

Intestinal microbiome is known to have contributory role in neonatal neurological growth because it can regulate energy generated from nutrition and it might also play a regulatory function in the improvement of impaired blood-brain barrier as well as myelination, which are crucial microbiota roles during functional and structural neurodevelopment (Hoban et al., 2016).

Robertson and colleagues claimed that some bacterial signatures are linked with either malnutrition as well as a healthy development in the first 1,000 days. On the other hand, a variety of microbiomes with enhanced *Gemella spp., Corynebacterium, and Prevotella spp.* have been connected to decreases in length-for-age Z-score for newborn. Healthy growth is linked with higher amount of *Streptococcus thermophilus* and *Bifidobacterium longum* within the first 6 months but less predominant in neonatal malnutrition. The authors also reported that in late childhood, the substantial prevalence of some microbiota such as *Akkermansia muciniphila*, *Methanobrevibacter smithiiobligate* anaerobes, and *Lactobacillus* was linked to healthy development but *Staphylococcus aureus*, *Escherichia coli*, and other species are linked with severe undernutrition (Robertson et al., 2019).

There is always an indirect or direct effect of microbiota on the fetus of mammals either through immunological responses of the mothers, factors, or microbial metabolic products crossing the placenta membrane (Romano-Keeler and Weitkamp, 2015). It was recorded also that indirectly some factors like neuroendocrine exposure, stress, and diet mediate fetus epigenetic programming (Weaver et al., 2017; Yu et al., 2018) and affect maternal microbiome (Hoffman et al., 2019). Also, changes occur during gestation for both the vaginal (Aagaard et al., 2012) and intestinal microbiome (Koren et al., 2012). Intestinal microbiome was reported to assist fetal efficient energy acquisition from the mother through the placenta during pregnancy development (Koren et al., 2012), sustain integrity and function of intestinal epithelial, and support mother's immune tolerance through the microbiome product butyrate (Blaser and Dominguez-Bello, 2016).

Dominguez-Bello et al. (2019) described that pregnant women and infants are prone to infections by the microbiota due to exposure to complex microbiota. As a result, the primordial microbiota is inherited from mother to the newly born infants and thus affects the weight. It is worth noting that animal development requires microbiota but records have shown that microbiota germ-free mice have the abnormal phenotype. The growth of the metabolic, hormonal, immunological, microbiota, and neurological systems in infants have been linked to bacteria (Smith et al., 2015). Also, neonate and microbiota development is dependent on prebiotic, hormonal, nutritional, and immunological factors of maternal milk (Pacheco et al., 2015). However, during labor, mothers acquire some bacteria like lactate bacteria which digest lactose, and some indigestible substrates like glycans in milk known as human milk oligosaccharides (Hamilton et al., 2017) which help in shaping the newborn gut microbiota and breast milk microbial constituents mostly *Bifidobacterium* species (Garrido et al., 2013).

This strict lactation causes drastic development of both the motor and sensor neurons in the infant's brain. The microbiome is also involved in development like in the case of vitamin synthesis during postnatal growth (Dominguez-Bello et al., 2019). This is because fetal vitamin K diffusion via the placental barrier is poor (Lippi and Franchini, 2011) and thus cause a low amount of vitamin K in the neonate (Van Winckel et al., 2009). Consequently, there is an observed decrease in gut alpha diversity in the neonate (Bokulich et al., 2016) which could result from selective pressure in milk substrate and after 7 days of age causing similarity in intestinal microbiome development of 1 week with that of 1-month-old baby (Hill et al., 2017). However, neonate development in the first 6 months under specific stress of breast milk reshapes the intestinal microbiome community with metabolic products promoting production of peripheral regulatory T cell (Arpaia et al., 2013).

6.2.1 Diet-Microbiome-Epigenetics Crosstalk

Diets have the ability to influence the development of health and diseases. Dietary interventions or manipulations have been employed for the management of disease conditions, e.g., ketogenic diet for management of epilepsy. Also diets can also precipitate disease pathogenesis. However, a pertinent question is how is dietary supplementation able to influence human physiologic process? One of the main reasons for this effect is the diet-microbiome crosstalk. Diet is capable of reshaping the gut microbiome by changing the abundance of the commensal microbes or altering the gut microbial homeostasis resulting in dysbiosis. Low microbial diversity and dysbiosis have modulatory activity on the amounts of microbiota-derived nutrients, and metabolic products can serve as epigenetic modifiers through histone modifications and regulation of DNA methylation. Diet-mediated epigenetic alterations occur through the interference with the genomic enzymatic machinery through direct inhibition of the enzymes or modification of substrates availability (Poirier et al., 2001). Therefore, diet-microbiome-epigenetic crosstalk plays a significant role in the regulation of physiological activities and underscores the importance of dietary intervention for the modulation of microbiome and regulation of healthy lifestyle (Miro-Blanch and Yanes, 2019).

The involvement of diet in reshaping the microbiome and its fundamental function in coordinating the host-microbiome crosstalk is evident in the developmental phase of life, when breastfeeding neonates enhanced intestinal microbiota development (Charbonneau et al., 2016; Miro-Blanch and Yanes, 2019). Subsequently there is improved abundance of commensal microbes due to the consumption of solid foods (Laursen et al., 2017), and finally reduced proliferation of these commensal microbes due to aging process in the elderly which may be because of the low food diversity (Claesson et al., 2012). The intestinal microbial flora are sensitive to quantities of some dietary nutrients (Turnbaugh et al., 2009), and they also react in a different way to diet in a numerous environmental settings.

Epigenetic modulations are part of the major mechanistic approach through which environmental factors can regulate early stage intracellular growth and maturation. This produces a distinct phenotypic expression in pregnancy and during early infancy without any alteration of the DNA sequence. There are various factors that affect human development during intrauterine and neonatal stages. These include configuration/constitution of gut microbiome, pollutant/xenobiotic exposure, and maternal and neonatal diets (Indrio et al., 2017).

All of these factors have the capability to influence the formation of epigenetic alterations that play significant roles in the regulation of human survival. Furthermore, they can also modify physiologic processes for the maintenance of health and diseases through manipulations of the immunological response and inflammatory cytokines (Indrio et al., 2017). Microbiota colonization following delivery is determined by the following: delivery mode, maternal microflora, neonatal diets, and early skin contacts resulting in epigenetic modulations that gut microbiome constitution and function especially their responses to challenges or health assaults (Indrio et al., 2017).

In addition, epigenetic influences are most effective and strongly impressed on the DNA within the windows of conceptions and the second year birthday which is the first 1,000 days of life of the baby (Bhutta et al., 2013). Developmental programming of metabolic and cardiovascular diseases that affect the individual later in life begins during these 1,000 days of life when epigenetic signatures are mostly potent and impressed on the DNA. Interestingly, it has been reported that there is transgenerational inheritance of some of these epigenetic alterations caused by neonatal dietary supplementation and gut microbiota interactions resulting in major evolutionary changes (Miller, 2016).

The most interesting diet-microbiota-epigenetic interactions are associated with the production of numerous metabolic products such as lipids, vitamins, and bile acids (Nicholson et al., 2012) that find their way into the host circulatory system. Also colonic microbiome fermentation of indigestible carbohydrates or fibers in the colon results in production of metabolites which include butyrate, propionate, and acetate generally called SCFAs (den Besten et al., 2013). SCFAs are linked to disease pathogenesis and some beneficial effects have also been suggested (Belcheva et al., 2014; Canfora et al., 2015; Perry et al., 2016), provoking necessary enhanced elucidation of their mechanistic pathways (Figure 6.1).

Microbial dietary metabolism generates compounds that modify epigenetics (Davis and Ross, 2007). Methylation of DNA is by far the most explored epigenetics alteration. There are foods that have the capacity to modify methylation of the DNA. The following foods or nutrients have DNA methylation properties: turmeric, citruses, tomatoes, broccoli, cinnamon, apples, soybean, coffee, tea, etc. (Huang et al., 2019).

SCFAs are known epigenetic modulators which regulate host genetic manifestation through inhibition of histone deacetylase (HDAC) (Olaniyi et al., 2020). Sodium acetate has been reported to improve diabetes-induced cardiac dysfunction and hepatic lipid dysregulation through the inhibition of HDAC activity (Olaniyi et al., 2020; Olaniyi and Amusa, 2020). The gut microbiome is a vast store house for generating large quantities of diverse metabolites that have possible HDAC modulatory actions and other epigenetic modifications (Figure 6.1; Table 6.1).

6.3 MICROBIOME, IMMUNITY, AND EPIGENETIC MODIFICATION

Numerous mental illnesses like schizophrenia have multifaceted pathology. It has been affirmed that the initiation and the development of diseases might not be described by one-disease and one-gene techniques. It has been observed that the contribution of numerous genes might play a significant function in the level of their pathobiology. Most of these interactions are possibly controlled by epigenetics modulating machineries. When likened to genetic mutations, it has been observed that the alternation in epigenetic machinery such as RNA interference, histone modifications, and DNA methylation could impact gene manifestation. Hence this might have resulted into neuronal and behavioral alteration discovered in mental disorders. It has

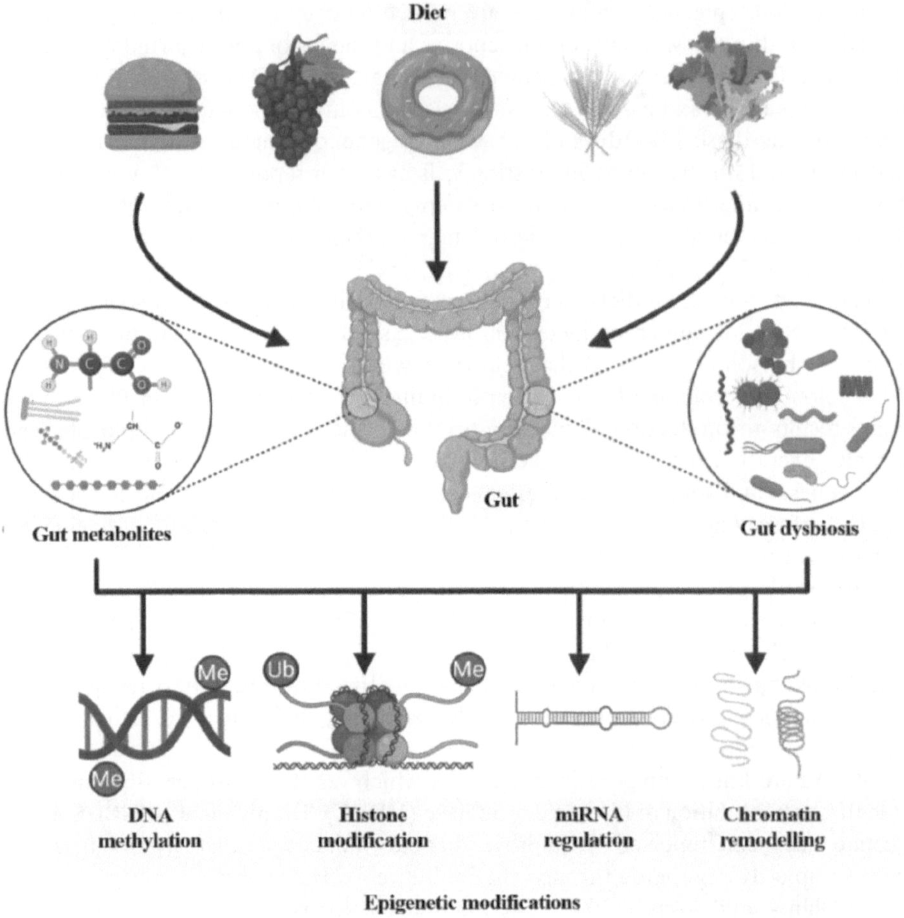

FIGURE 6.1 Diet-microbiome-epigenetic interactions.

been observed that influences from the environment like infections, intestinal microbiome, and diet exhibit noteworthy influence in the observed epigenomic modifications. Investigation revealed that the gut microbiota and bioactive nutrients portend the potential to change histone signatures and DNA methylation through numerous modes of action. Certainly, it has been observed (Stilling et al., 2014; Mayer et al., 2014; Kelly et al., 2015; Dinan and Cryan, 2017) that microorganisms available inside the intestinal mucosal could perform momentous function of stimulating numerous elements of gut-brain axis, which represents crosstalk between gastrointestinal tract and the central nervous system. This gut-brain relationship was further demystified by Alam and coworkers (2017) who demonstrated how dietary modulation results in alteration in intestinal microbiota causing inflammatory and epigenetic changes noticed in neuropsychiatric diseases. Therefore, the gut and brain actually interact through the microbiome-mediated biochemical signaling (Stilling et al., 2014; Mayer et al., 2014; Kelly et al., 2015; Dinan and Cryan, 2017).

TABLE 6.1
Studies Showing Microbiome Reshaping and Epigenetic Modification

S/N	Changes in Microbiota	Epigenetic Modifications	Effect	References
1	Maternal diets, e.g., high-fat diet	Reprogramed DNA methylation, altered histone acetylation, and methylation (impact fetal as well as neonatal microbiome establishment)	Elevated fetal birth weight and predisposition to obesity later in adult	Li (2018), Parlee and MacDougald (2014)
2	Maternal under nutrition	MicroRNA changes caused alteration in histone and DNA methylation (altered fetal microbiome)	Heightened susceptibility to impaired glucose regulation, reduced insulin sensitivity, and overt diabetes in the offspring	Parlee and MacDougald (2014), Li et al. (2010)
3	Sodium acetate (SCFA) supplementation in animal model of diabetes	Inhibition of histone deacetylase activity	Improve diabetes-induced cardiac dysfunction and hepatic lipid dysregulation	Olaniyi et al. (2020), Olaniyi and Amusa (2020)
4	High fat diet-induced obesity (altered gut microbiome)	Dysregulation of the expression of microRNAs (MIR-181)	Dysmetabolism and white adipose tissue inflammation resulting in obesity and associated metabolic disorders	Virtue et al. (2019)
5	Obesogenic diet shapes the microbiome	Alterations of histone methylation and acetylation	Reprogramming of the intestinal epigenomic machinery resulting impaired gene expression causing colonic cancer	Qin et al. (2018)

(*Continued*)

TABLE 6.1 (*Continued*)
Studies Showing Microbiome Reshaping and Epigenetic Modification

S/N	Changes in Microbiota	Epigenetic Modifications	Effect	References
6	Vitamin D-deficient diet in pregnancy	Methylation of DNA in the placenta	Pathological pregnancy, e.g., preeclampsia and adverse fetal outcome	Anderson et al. (2015)
7	Choline-rich diet in pregnancy	Histone and DNA methylation	Enhanced brain development and functioning	Davison et al. (2009)
8	L-carnitine dietary supplementation	Enhanced microbial synthesis of TMA/TMAO	Increased CVD risk, e.g., atherosclerosis	Koeth et al. (2013)
9	Consumption of green tea polyphenols, e.g., Epigallocatechin-3-gallate	Suppress DNA methylation by inhibiting DNA methyltransferases resulting in demethylation and reactivation of genes	Reduced oxidative damage and angiogenesis, inhibition of proliferation of cancerous cells, initiation of apoptosis	Fang et al. (2003), Lee et al. (2005), Yang et al. (2009)
10	Microbiota dysbiosis–induced blood-brain barrier dysfunction	Neurotransmitters and gut microbial metabolites such as vitamins, lipids, L-carnitine, phosphatidylcholine, glutamate, trimethylamine (TMAO), bile acids	Alzheimer's disease, Schizophrenia, autism spectrum disorders (ASDs), Parkinson's disease, and anxiety and depression	Winek et al. (2016), Fiorentino et al. (2016), Zhu et al. (2020)

Environmental or nongenetic factors with epigenetic regulating function have been reported as novel initiating elements for the pathogenesis of eye disease through intraocular inflammation-related mechanism (Wen et al., 2018). Microbiota that colonize the ocular surface have been linked to intraocular inflammation through epigenetic regulatory processes. Uncovering of the intraocular microbiome through high-throughput DNA sequencing technique represents a game changer in the demystifying the interaction between microbiome and ocular diseases such as uveitis and age-related macular degeneration. Since microbial dysbiosis and microbiota-metabolites have been linked to the pathobiology of inflammation-mediated ocular disorders, analysis of the intraocular microbiome is definitely a game changer in the management of ocular diseases (Wen et al., 2018).

Metabolic diseases have been discovered to trigger several alarming health concerns. Several environmental factors have been shown to be connected with numerous metabolic disorders and they perform a significant function during the process (Portha et al., 2014; Tilg et al., 2020). Diet is known to play crucial role in the modulation of metabolic dysfunctions through alterations of the gut microbiota. Microbiome exerts significant functions in numerous metabolic activities most especially in humans and could be a great factor that regulates host metabolism. The high rate of alterations in the roles and configuration of the intestinal microbiome exhibits vital functions in the pathobiology of metabolic illness through the stimulation of the epigenetic alteration such as histone alterations, noncoding RNAs, and DNA methylation. Moreover, it has also been discovered that the process involved in the induction of epigenetic alteration could be controlled through the action of metabolites generated by intestinal microbiome entailing SCFAs, trimethylamine-N-oxide, and folates. Moreover, investigations have affirmed the significant function of these microbial-generated metabolites most especially in the pathophysiology of diabetes and obesity. Therefore, Sharma et al. (2019) investigated the associations that exist between the epigenetics and gut microbiome in the control and developmental processes involved in diabetes and obesity. The gut microbiome is under the regulatory activities of diets, genetics, and epigenetics. Epigenetic regulation of the intestinal microbiome especially in obese and diabetic conditions is beginning to gain deserved research attention, resulting in unraveling of innovative mechanistic target for the reinstatement of gut microbiota homeostasis for the prevention, treatment, or management of diabetes and obesity.

The application of omics-based technologies has been developed for over a period of time, and the application of relevant modes of action through microbiomes could influence host physiologic processes. The multi-omics approach has been used for the extensive sequencing of the microbiome which has broad influence on the host metabolism, genetics, and epigenetics resulting in the modification regulatory mechanisms involved the maintenance of health and disease pathogenesis (Wang et al., 2019).

Gut microbiome portends the capacity to give necessary environmental insight on how to alter the host microbiome rejoinders through changes in the epigenomic composition and eventually genetic manifestation. Microbiota generated metabolites have been linked with genomic and epigenomic transcription processes which alters the gene expression. Qin and Wade (2018) provide detailed information on the proper facts of the host-microbiome-epigenetic which entails DNA methylation. Several trillion microbes colonize the intestine of humans. Technological advances using next-generation genome sequencing procedures is beginning to unravel the unique characteristics and roles of the intestinal microbial flora. Intestinal microbial flora have been shown to have interesting communication with the various organ systems of the body with exciting coordinating linkages that influence physiological functions and pathophysiological processes in the host through microbiome-nutrient-epigenetic crosstalk (Ursell et al., 2012).

Zhang et al. (2019) stated the recent facts on the activity of the intestinal microbiota in metabolism, immunity, microbiome influence on extraintestinal cancer as well as the immune rejoinder, and stratagems to control the gut microbiome. It has highlighted that the mammalian intestine entails significant number of microorganisms as well as beneficial metabolites which constitutes basis for the stimulation of the immunological functions of the host. It has been observed that the imbalance among commensal organisms and immunological system of the gut interrupts intestinal microbiological homeostasis that could result into microbiota dysbiosis, proinflammatory immune rejoinders, impaired reliability of the gut barrier, toward symbionts.

6.3.1 Microbiota, Epigenetics, and Neurodegeneration

The gut microbiota, an epigenetic entity, weighs about 1–2 kg in adult human body similar to the weight of the brain (Parent and Carpenter, 1996; Forsythe and Kunze, 2013; Stilling et al., 2014). There is a continuous relationship between mammals and the environmental microorganisms which eventually colonize the gut of mammals such that they both undergo growth and evolution together. The most significant thing about microbiome-host interactions is that it is mutually beneficial because there is interdependency for survival and other vital functions (Bordenstein and Theis, 2015). The intestinal microbes are implicated in regulating brain growth in neonates (Bercik et al., 2012; Collins et al., 2012; Mayer et al., 2014; Sampson and Mazmanian, 2015). Microglia, immune cells, of the brain has been implicated in neuroinflammation and aging (Jyothi et al., 2015). Interestingly, intestinal microbial flora have been consistently linked with microglia regulation (Erny et al., 2015). Hence, the neuroimmunological activities can be modulated by altering the gut microbiota. SCFAs have been implicated in these neuroimmune responses of the intestinal microbiomes.

Gut microbiota and its metabolic products have capability to penetrate the blood-brain barrier (BBB), as such; they have consequences on some neurodegenerative disorders associated with impaired/leaky BBB (Braniste et al., 2014). Furthermore, stress has serious influence on the gut microbiome (Moloney et al., 2014) which may be detrimental to aging brain. Therefore, aging and stress influence the gut microbiome to initiate pro-inflammation and impaired gut barrier function (Kelly et al., 2015). Similarly, stress can also have harmful effects on integrity of the BBB especially the aging brain (Montagne et al., 2015; Esposito et al., 2002). Therefore, impaired microbiome homeostasis and BBB integrity have both been linked to the progression of neurodegenerative tendencies in the brain. Indeed, detailed elucidation of the mechanistic processes associated with the microbiome-brain interactions especially in the pathobiology of neurodegeneration is noteworthy.

Deregulation of intestinal microbiota homeostasis is associated with numerous diseases (Diaz Heijtz et al., 2011; Dinan and Cryan, 2017). The microbiome-brain crosstalk contributory role in the development of neurological disorders is progressively being acknowledged (Smith, 2015; Bell et al., 2019). Microbiome transplantation revealed microbial flora interactions with neurobehavior as reported by Bercik et al. (2011). Furthermore, germ-free rodents were reported to demonstrate a low anxiety-like neurobehavioral phenotype (Diaz Heijtz et al., 2011). Likewise, microbiota was linked to Alzheimer's disease as depicted by the reduction in neuroinflammation as well as cerebral Amyloid β pathology (Harach et al., 2017; Jiang et al., 2017).

Epigenetics drive inheritable modifications in genetic manifestation. It involved the mechanistic process by which previous events result in alteration of the synaptic plasticity, memory attainment, and merging neural circuits connection and information flow. Therefore, alteration of chromatin can permit modifications in the genes expression connected to plasticity of synapse during growth and in adult (Fagiolini et al., 2009), and hence permit genetic susceptibility, and neural function synergize to change neurobehavioral and cell rejoinders. Additionally, epigenetic deregulation has been linked to cognitive decline and neural cell death connected with neurodegenerative diseases (Hwang et al., 2013; Day et al., 2015; Landgrave-Gomez et al., 2015). Specifically, numerous diseases with diverse root causes and/or pathogenesis, cognitive decline, and neural cell death involve epigenetic alteration of genes.

6.4 FUTURE PERSPECTIVES

The enormous influence of the gut microbiome on human health cannot be overestimated. Given the gap in knowledge, more concerted, integrated, clinical, and multicenter randomized controlled studies need to be carried out to provide more clinical evidences that can be translated from the bench to the bedside especially in Africa. The microbiota-diet-epigenetics axis should be given further attention because it has immense potentials for the treatment and prevention of chronic metabolic disorders, immunological dysfunctions, neurological diseases, etc. Fecal microbiome transplantation (FMT) is a very good example of the benefits of microbiome research in disease management or treatment. FMT was recently demonstrated clinically to restore gut microbiome of newborn delivered through cesarean section to the normal level seen in vaginal delivered infants through introduction of the maternal microbes (Korpela et al., 2020). Microbiome should be targeted for therapeutic and pharmacological management of disease conditions because microbiome is an epigenetic entity on its own. Microbiota-targeted drugs will reveal the significant epigenetic contributions of the microbiome. Microbiome-mediated transgenerational inheritance of metabolic disorders can be reduced through the targeting and/or reshaping of microbiome during the developmental window to reprogram the fetus of obese or diabetic mothers, thereby preventing inheritance of metabolic disorders. Furthermore, microbiota-based dietary and probiotic supplements may become indispensable for personalized care and dietary preparation for efficient therapy. Fecal microbiome profiling may soon become a routine medical examination for the treatment of diseases. Incorporation of microbiome-epigenetics interactions into clinical care will expose therapeutic and mechanistic application of clinical epigenetics for the translational management of debilitating disorders like immunological, neurodegenerative, metabolic, and cardiovascular diseases. In essence, microbiome could become a diagnostic instrument for the detection, treatment, and management of diseases using cutting-edge techniques like whole genome sequencing, metagenomic, and metabolomic technologies.

6.5 CONCLUSION

Microbiome-mediated regulations of host reaction to environmental influences are achieved through the alterations of the host epigenomic composition without affecting the DNA sequences. Microbial metabolites have been identified as potent epigenetic modifiers. These metabolites cause histone modification, DNA methylation, and RNA-associated changes (Figure 6.2). These epigenomic alterations result in the pathogenesis of hypertension, cancer, neurodegeneration, etc. Interestingly, the

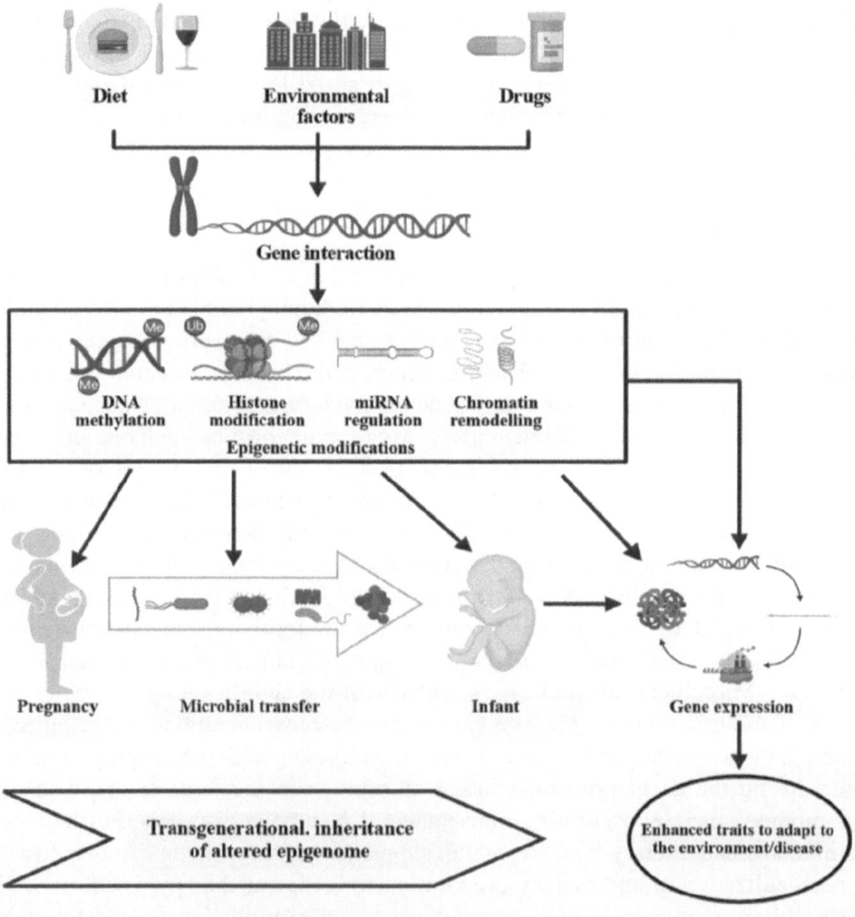

FIGURE 6.2 Microbiome reshaping, epigenetic modification, and transgenerational inheritance.

altered genetic expressions are inheritable, meaning that there can be transgenerational transmission of the genetic modifications induced by microbiome-epigenetic interactions. The microbiome-epigenetic signaling and communications can provide interesting insight or pharmacologic target for the management of some debilitating disease conditions like Alzheimer's diseases, Parkinson's disease, diabetes mellitus, hypertension, etc. Therefore, there is need for further clarification of the host-microbiome-epigenetic interactions in health and diseases.

REFERENCES

Aagaard K., Ma J., Antony K.M., Ganu R., Petrosino J., Versalovic J. (2014) The placenta harbors a unique microbiome. *Science Translational Medicine*, 6(237): 237ra65. doi: 10.1126/scitranslmed.3008599.

Aagaard K., Riehle K., Ma J., Segata N., Mistretta T.A., Coarfa C., Raza S., Rosenbaum S., Van den Veyver I., Milosavljevic A., Gevers D., Huttenhower C., Petrosino J., Versalovic J. (2012). A metagenomic approach to characterization of the vaginal microbiome signature in pregnancy. *PLoS One*, 7(6): e36466. https://doi.org/10.1371/journal.pone.0036466.

Adeyanju O.A., Badejogbin O.C., Areola D.E., Olaniyi K.S., Dibia C., Soetan O.A., Oniyide A.A., Michael O.S., Olatunji L.A., Soladoye A.O. (2020) Sodium butyrate arrests pancreato-hepatic synchronous uric acid and lipid dysmetabolism in high fat diet fed Wistar rats. *Biomedicine and Pharmacotherapy*, 133: 110994. doi: 10.1016/j.biopha.2020.110994.

Alam R., Abdolmaleky H.M., Zhou J. (2017) Microbiome, inflammation, epigenetic alterations, and mental diseases. *American Journal of Medical Genetics, Part B: Neuropsychiatric Genetics*, 174(3). doi: 10.1002/ajmg.b.32567.

Allin K.H., Nielsen T., Pedersen O. (2015) Mechanisms in endocrinology: Gut microbiota in patients with type 2 diabetes mellitus. *European Journal of Endocrinology*, 172(4): R167–R177. doi: 10.1530/eje-14-0874.

Alvarez-Venegas R. (2014) Bacterial SET domain proteins and their role in eukaryotic chromatin modification. *Frontiers in Genetics*, 5: 1–8.

Anderson C.M., Ralph J.L., Johnson L., et al. (2015) First trimester vitamin D status and placental epigenomics in preeclampsia among Northern Plains primiparas. *Life Science*, 129. doi: 10-510.1016/j.lfs.2014.07.012.

Arpaia N., Campbell C., Fan X., Dikiy S., van der Veeken J., deRoos P., Liu H., Cross J.R., Pfeffer K., Coffer P.J., Rudensky A.Y.(2013) Metabolites produced by commensal bacteria promote peripheral regulatory T-cell generation. *Nature*, 504(7480):451–455.

Asnicar F., Manara S., Zolfo M., Truong D.T., Scholz M., Armanini F., Ferretti P., Gorfer V., Pedrotti A., Tett A., Segata N. (2018) Studying vertical microbiome transmission from mothers to infants by strain-level metagenomic profiling. *mSystems*, 2: e00164–16.

Belcheva A., Irrazabal T., Robertson S.J., et al. (2014) Gut microbial metabolism drives transformation of MSH2-deficient colon epithelial cells. *Cell*, 158: 288–299.

Bell J.S., Spencer J.I., Yates R.L., Yee S.A., Jacobs B.M., DeLuca G.C. (2018) From nose to gut – The role of the microbiome in neurological disease. *Neuropathology and Applied Neurobiology*, 45(3): 195–215.

Bendiks M., Kopp M.V. (2013) The relationship between advances in understanding the microbiome and the maturing hygiene hypothesis. *Current Allergy and Asthma Reports*, 13: 487–494.

Bercik P., Collins S.M., Verdu E.F. (2012) Microbes and the gut-brain axis. *Neurogastroenterology and Motility*, 24: 405–413.

Bercik P., Denou E., Collins J., et al. (2011) The intestinal microbiota affect central levels of brain-derived neurotropic factor and behavior in mice. *Gastroenterology*, 141: 599–609.
Bhutta Z.A., Das J.K., Rizvi A., et al. (2013) Evidence based interventions for improvement of maternal and child nutrition: What can be done and at what cost? *Lancet*, 382: 452–477. doi: 10.1016/ S0140-6736(13)60996-4.
Bierne H., Hamon M. (2020) Targeting host epigenetic machinery: The Listeria paradigm. *Cell Microbiol*, 22(4): e13169.
Blaser M.J., Dominguez-Bello, M.G. (2016) The human microbiome before birth. *Cell Host and Microbe*, 20(5): 558–560. https://doi.org/10.1016/j.chom.2016.10.014.
Bokulich N.A., Chung J., Battaglia T., et al. (2016) Antibiotics, birth mode, and diet shape microbiome maturation during early life. *Science Translational Medicine*, 8(343): 343ra82. doi: 10.1126/scitranslm ed.aad7121.
Bordenstein S.R., Theis K.R. (2015) Host biology in light of the microbiome: Ten principles of holobionts and hologenomes. *PLoS Biology*, 13: e1002226.
Braniste V., Al-Asmakh M., Kowal C., et al. (2014) The gut microbiota influences blood-brain barrier permeability in mice. *Science Translational Medicine*, 6: 263ra158. doi: 10.1126/scitranslmed.3009759.
Callahan B.J., DiGiulio D.B., Goltsman D., et al. (2017) Replication and refinement of a vaginal microbial signature of preterm birth in two racially distinct cohorts of US women. *Proceedings of the National Academy of Science USA*, 114(37): 9966–9971. doi: 10.1073/pnas.1705899114.
Canfora E.E., Jocken J.W., Blaak E.E. (2015) Short-chain fatty acids in control of body weight and insulin sensitivity. *Nature Reviews Endocrinology*, 11(10): 577–591. doi: 10.1038/ nrendo.2015.128.
Charbonneau M.R., O'Donnell D., Blanton L.V., et al. (2016) Sialylated milk oligosaccharides promote microbiota-dependent growth in models of infant undernutrition. *Cell*, 164(5): 859–871. doi: 10.1016/j.cell.2016.01.024.
Chiu C.Y., Chan Y.L., Tsai Y.S., Chen S.A., Wang C.J., Chen K.F., Chung I.F. (2017) Airway microbial diversity is inversely associated with mite-sensitized rhinitis and asthma in early childhood. *Scientific Reports*, 7: 1820. https://doi.org/10.1038/s41598-017-02067-7.
Chu D.M., Ma J., Prince A.L., Antony K.M., Seferovic M.D., Aagaard K.M. (2017) Maturation of the infant microbiome community structure and function across multiple body sites and in relation to mode of delivery. *Nature Medicine*, 23(3): 314–326. https://doi.org/10.1038/nm.4272.
Claesson M.J., Jeffery I.B., Conde S., et al. (2012) Gut microbiota composition correlates with diet and health in the elderly. *Nature*, 488(7410): 178–184.
Cobb C.M., Kelly P.J., Williams K.B., Babbar S., Angolkar M., Derman R.J. (2017) The oral microbiome and adverse pregnancy outcomes. *International Journal of Women's Health*, 9: 551–559.
Collins S.M., Surette M., Bercik P. (2012) The interplay between the intestinal microbiota and the brain. *Nature Reviews Microbiology*, 10(11): 735–742. doi: 10.1038/nrmicro2876.
Costabile A., Klinder A., Fava F., Napolitano A., Fogliano V., Leonard C., Gibson G.R., Tuohy K.M. (2008) Whole-grain wheat breakfast cereal has a prebiotic effect on the human gut microbiota: A double-blind, placebo-controlled, crossover study. *British Journal of Nutrition*, 99(1): 110–120. doi: 10.1017/S0007114507793923.
Cotillard A., Kennedy S.P., Kong L.C. (2013) Dietary intervention impact on gut microbial gene richness. *Nature*, 500(7464): 585–588. doi: 10.1038/nature12480.
Dahl C., Stanislawski M., Iszatt N., Mandal S., Lozupone C., Clemente J.C., Knight R., Stigum H., Eggesbø M. (2017) Gut microbiome of mothers delivering prematurely shows reduced diversity and lower relative abundance of Bifidobacterium and Streptococcus. *PLoS One*, 12(10): e0184336. https://doi.org/10.1371/journal.pone.0184336.

Dasanayake A.P., Li Y., Wiener H., Ruby J.D., Lee M.J. (2005) Salivary Actinomyces naeslundii genospecies 2 and Lactobacillus casei levels predict pregnancy outcomes. *Journal of Periodontology*, 76(2): 171–177.

Davidson S.K., Stahl D.A. (2006) Transmission of nephridial bacteria of the earthworm Eiseniafetida. *Applied and Environmental Microbiology*, 72(1): 769–775.

Davis C.D., Ross S.A. (2007) Dietary components impact histone modifications and cancer risk. *Nutrition Reviews*, 65(2): 88–94. doi: 10.1111/j.1753–4887.2007.tb00285.x.

Davison J.M., Mellott T.J., Kovacheva V.P., Blusztajn J.K. (2009) Gestational choline supply regulates methylation of histone H3, expression of histone methyltransferases G9a (Kmt1c) and Suv39h1 (Kmt1a), and DNA methylation of their genes in rat fetal liver and brain. *Journal of Biological Chemistry*, 284: 1982–1989. doi: 10.1074/jbc.M807651200.

Day J.J., Kennedy A.J., Sweatt J.D. (2015) DNA methylation and its implications and accessibility for neuropsychiatric therapeutics. *Annual Review of Pharmacology and Toxicology*, 55: 591–611. doi: 10.1146/annurev-pharmtox-010814-124527.

De Filippo C., Cavalieri D., Di Paola M., Ramazzotti M., Poullet J.B., Massart S., Collini S., Pieraccini G., Lionetti P. (2010) Impact of diet in shaping gut microbiota revealed by a comparative study in children from Europe and rural Africa. *Proceedings of the National Academy of Science, USA*, 107(33): 14691–14696. doi: 10.1073/pnas.1005963107.

den Besten G., van Eunen K., Groen A.K., Venema K., Reijngoud D.J., Bakker B.M. (2013) The role of short-chain fatty acids in the interplay between diet, gut microbiota, and host energy metabolism. *Journal of Lipid Research*, 54(9): 2325–2340.

Diaz Heijtz R., Wang S., Anuar F., Qian Y., Björkholm B., Samuelsson A., Hibberd M.L., Forssberg H., Pettersson S. (2011) Normal gut microbiota modulates brain development and behavior. *Proceedings of the National Academy of Science*, 108(7): 3047–3052.

Dinan T.G., Cryan J.F. (2017) Gut instincts: Microbiota as a key regulator of brain development, ageing and neurodegeneration. *Journal of Physiology*, 595(2): 489–503.

Dominguez-Bello M.G., Costello E.K., Contreras M., Magris M., Hidalgo G., Fierer N., Knight R. (2010) Delivery mode shapes the acquisition and structure of the initial microbiota across multiple body habitats in newborns. *Proceedings of the National Academy of Science USA*, 107(26): 11971–11975. doi: 10.1073/pnas.1002601107.

Dominguez-Bello M.G., Godoy-Vitorino F., Knight R., Blaser M.J. (2019) Role of the microbiome in human development. *Gut*, 68(6): 1108–1114.

Doyle R.M., Harris K., Kamiza S., Harjunmaa U., Ashorn U., Nkhoma M., Dewey K.G., Maleta K., Ashorn P., Klein N. (2017) Bacterial communities found in placental tissues are associated with severe chorioamnionitis and adverse birth outcomes. *PLoS One*, 12(7): e0180167.

Duncan S.H., Belenguer A., Holtrop G., Johnstone A.M., Flint H.J., Lobley G.E. (2007) Reduced dietary intake of carbohydrates by obese subjects results in decreased concentrations of butyrate and butyrate-producing bacteria in feces. *Applied Environmental Microbiology*, 73(4): 1073–1078.

Erny D., Hrabe de Angelis A.L., Jaitin D., et al. (2015) Host microbiota constantly control maturation and function of microglia in the CNS. *Nature Neuroscience*, 18(7): 965–977.

Esposito P., Chandler N., Kandere K., Basu S., Jacobson S., Connolly R., Tutor D., Theoharides T.C. (2002) Corticotropin-releasing hormone and brain mast cells regulate blood-brain-barrier permeability induced by acute stress. *Journal of Pharmacology and Experimental Therapeutics*, 303: 1061–1066.

Fagiolini M., Jensen C.L., Champagne F.A. (2009) Epigenetic influences on brain development and plasticity. *Current Opinion in Neurobiology*, 19: 207–212.

Fang M.Z., Wang Y., Ai N., Hou Z., Sun Y., Lu H., Welsh W., Yang C.S. (2003) Tea polyphenol (-)-epigallocatechin-3-gallate inhibits DNA methyltransferase and reactivates methylation-silenced genes in cancer cell lines. *Cancer Research*, 63(22): 7563–7570.

Fardini Y., Chung P., Dumm R., Joshi N., Han Y.W. (2010) Transmission of diverse oral bacteria to murine placenta: evidence for the oral microbiome as a potential source of intrauterine infection. *Infection and Immunity*, 78(4): 1789–1796

Fiorentino M., Sapone A., Senger S., Camhi S.S., Kadzielski S.M., Buie T.M., Kelly D.L., Cascella N., Fasano A. (2016) Blood-brain barrier and intestinal epithelial barrier alterations in autism spectrum disorders. *Molecular Autism*, 7: 49.

Fischer N. (2020) Infection-induced epigenetic changes and their impact on the pathogenesis of diseases. *Seminars in Immunopathology*, 42(2): 127–130.

Fol M., Włodarczyk M., Druszczyńska M. (2020) Host epigenetics in intracellular pathogen infections. *International Journal of Molecular Sciences*, 21(13): 4573.

Forsythe, P., Kunze, W.A. (2013) Voices from within: Gut microbes and the CNS. *Cellular and Molecular Life Sciences*, 70: 55–69. doi: 10.1007/s00018-012-1028-z.

Garrido D., Ruiz-Moyano S., Jimenez-Espinoza R., Eom H.J., Block D.E., Mills D.A. (2013) Utilization of galactooligosaccharides by Bifidobacteriumlongum subsp. infantis isolates. *Food Microbiology*, 33(2): 262–270. doi: 10.1016/j.fm.2012.10.003.

Getahun D., Fassett M.J., Longstreth G.F., Koebnick C., Langer-Gould A.M., Strickland D., Jacobsen S.J. (2014) Association between maternal inflammatory bowel disease and adverse perinatal outcomes. *Journal of Perinatology*, 34: 435–440.

Gomez-Arango L.F., Barrett H.L., McIntyre H.D., Callaway L.K., Morrison M., Nitert M.D. (2017) Contributions of the maternal oral and gut microbiome to placental microbial colonization in overweight and obese pregnant women. Sci Rep, 7(1):2860. doi: 10.1038/s41598-017-03066-4.

Hamilton M.K., Ronveaux C.C., Rust B.M., Newman J.W., Hawley M., Barile D., Mills D.A., Raybould H.E. (2017) Prebiotic milk oligosaccharides prevent development of obese phenotype, impairment of gut permeability, and microbial dysbiosis in high fat-fed mice. *American Journal of Physiology, Gastrointestinal and Liver Physiology*, 312(5): G474–G487. doi: 10.1152/ajpgi.00427.2016.

Han M., Jia L., Lv W., Wang L., Cui W. (2019) Epigenetic enzyme mutations: Role in tumorigenesis and molecular inhibitors. *Frontiers in Oncology*, 29(9): 194.

Harach T., Marungruang N., Duthilleul N., Cheatham V., Mc Coy K.D., Frisoni G., Neher J.J., Fåk F., Jucker M., Lasser T., Bolmont T. (2017) Reduction of Abeta amyloid pathology in APPPS1 transgenic mice in the absence of gut microbiota. *Scientific Reports*, 7: 41802. doi: 10.1038/srep41802.

Hill C.J., Lynch D.B., Murphy K., Ulaszewska M., Jeffery I.B., O'Shea C.A., Watkins C., Dempsey E., Mattivi F., Tuohy K., Ross R.P., Ryan C.A., O'Toole P.W., Stanton C. (2017) Evolution of gut microbiota composition from birth to 24 weeks in the INFANTMET Cohort. *Microbiome*, 5(1): 4. doi: 10.1186/s40168-016-0213-y.

Hoban A.E., Stilling R.M., Ryan F.J., Shanahan F., Dinan T.G., Claesson M.J., Clarke G., Cryan J.F. (2016) Regulation of prefrontal cortex myelination by the microbiota. *Translational Psychiatry*, 6(4): e774. doi: 10.1038/tp.2016.42.

Hoffman J.D., Yanckello L.M., Chlipala G., Hammond T.C., McCulloch S.D., Parikh I., Sun S., Morganti J.M., Green S.J., Lin A.L.(2019) Dietary inulin alters the gut microbiome, enhances systemic metabolism and reduces neuroinflammation in an APOE4 mouse model. *PLoS ONE*, 14(8): e0221828.

Huang D., Cui L., Ahmed S., Zainab F., Wu Q., Wang X., Yuan Z. (2019) An overview of epigenetic agents and natural nutrition products targeting DNA methyltransferase, histone deacetylases and microRNAs. *Food Chemistry and Toxicology*, 123: 574–594. doi: 10.1016/j.fct.2018.10.052.

Huang J., Plass C., Gerhauser C. (2011) Cancer chemoprevention by targeting the epigenome. *Current Drug Targets*, 12: 1925–1956.

Hullar M.A., Burnett-Hartman A.N., Lampe J.W. (2014) Gut microbes, diet, and cancer. *Cancer Treatment and Research*, 159: 377–399.

Hullar M.A., Fu B.C. (2014) Diet, the gut microbiome, and epigenetics. *Cancer Journal (Sudbury, Mass.)*, 20(3): 170–175. doi: 10.1097/PPO.0000000000000053.

Hwang J.Y., Aromolaran K.A., Zukin R.S. (2013) Epigenetic mechanisms in stroke and epilepsy. *Neuropsychopharmacology*, 38: 167–182.

Hylla S., Gostner A., Dusel G., Anger H., Bartram H.P., Christl S.U., Kasper H., Scheppach W. (1998) Effects of resistant starch on the colon in healthy volunteers: Possible implications for cancer prevention. *American Journal of Clinical Nutrition*, 67(1): 136–142.

Indrio F., Martini S., Francavilla R., Corvaglia L., Cristofori F., Mastrolia S.A., Neu J., Rautava S., Russo Spena G., Raimondi F., Loverro G. (2017) Epigenetic matters: The link between early nutrition, microbiome, and long-term health development. *Frontiers in Pediatrics*, 5: 178. doi: 10.3389/fped.2017.00178.

Jiang C., Li G., Huang P., Liu Z., Zhao B. (2017) The gut microbiota and Alzheimer's disease. *Journal of Alzheimers Disease*, 58: 1–15.

Johnson S.K., Chua V., Hall R.S., et al. (2006) Lupin kernel fibre foods improve bowel function and beneficially modify some putative faecal risk factors for colon cancer in men. *British Journal of Nutrition*, 95: 372–378.

Jyothi H.J., Vidyadhara D.J., Mahadevan A., Philip M., Parmar S.K., Manohari S.G., Shankar S.K., Raju T.R., Alladi P.A. (2015) Aging causes morphological alterations in astrocytes and microglia in human substantia nigra pars compacta. *Neurobiology of Aging*, 36: 3321–3333.

Kelly T.N., Bazzano L.A., Ajami N.J., He H., Zhao J., Petrosino J.F., Correa A., He J. (2016) Gut microbiome associates with lifetime cardiovascular disease risk profile among Bogalusa Heart Study participants. *Circulation Research*, 119(8): 956–964.

Kelly J.R., Kennedy P.J., Cryan J.F., Dinan T.G., Clarke G., Hyland N.P. (2015) Breaking down the barriers: The gut microbiome, intestinal permeability and stress-related psychiatric disorders. *Frontiers in Cellular Neuroscience*, 9: 392.

Koeth R.A., Wang Z., Levison B.S., Buffa J.A., Org E., Sheehy B.T. (2013) Intestinal microbiota metabolism of lcarnitine, a nutrient in red meat, promotes atherosclerosis. *Nature Medicine*, 19(5): 576–585.

Koren O., Goodrich J.K., Cullender T.C., Spor A., Laitinen K., Bäckhed H.K., Gonzalez A., Werner J.J., Angenent L.T., Knight R., Bäckhed F., Isolauri E., Salminen S., Ley R.E. (2012) Host remodeling of the gut microbiome and metabolic changes during pregnancy. *Cell*, 150(3): 470–480. doi: 10.1016/j.cell.2012.07.008.

Korpela K., de Vos W.M. (2018) Early life colonization of the human gut: Microbes matter everywhere. *Current Opinion in Microbiology*, 44: 70–78.

Korpela K., Helve O., Kolho K.L., Saisto T., Skogberg K., Dikareva E., Stefanovic V., Salonen A., Andersson S., de Vos W.M. (2020) Maternal fecal microbiota transplantation in cesarean-born infants rapidly restores normal gut microbial development: A proof-of-concept study. *Cell*, 183: 1–11.

Landgrave-Gomez J., Mercado-Gomez O., Guevara-Guzman R. (2015) Epigenetic mechanisms in neurological and neurodegenerative diseases. *Frontiers in Cellular Neuroscience*, 9: 58.

Laursen M.F., Bahl M.I., Michaelsen K.F., Licht T.R. (2017) First foods and gut microbes. *Frontiers in Microbiology*, 8: 356.

Lee W.J., Shim J.Y., Zhu B.T. (2005) Mechanisms for the inhibition of DNA methyltransferases by tea catechins and bioflavonoids. *Molecular Pharmacology*, 68(4): 1018–1030.

Li Y. (2018) Epigenetic mechanisms link maternal diets and gut microbiome to obesity in the offspring. *Frontiers in Genetics*, 9: 342. doi: 10.3389/fgene.2018.00342.

Li C.C., Maloney C.A., Cropley J.E., Suter C.M. (2010) Epigenetic programming by maternal nutrition: Shaping future generations. *Epigenomics*, 2: 539–549. doi: 10.2217/epi.10.33.

Lin A., Bik E.M., Costello E.K., Dethlefsen L., Haque R., Relman D.A., Singh U. (2013) Distinct distal gut microbiome diversity and composition in healthy children from Bangladesh and the United States. *PLoS One*, 8(1): e53838. doi: 10.1371/journal.pone.0053838.

Lin J.C., Jeong S., Liang G., Takai D., Fatemi M., Tsai Y.C., Egger G., Gal-Yam E.N., Jones P.A. (2007) Role of nucleosomal occupancy in the epigenetic silencing of the MLH1 CpG island. *Cancer Cell*, 12(5): 432–444.

Lippi G., Franchini M. (2011) Vitamin K in neonates: Facts and myths. *Blood Transfusion*, 9(1): 4–9. doi: 10.2450/2010.0034-10.

Marlicz W., Skonieczna-Zydecka K., Dabos K.J., Loniewski I., Koulaouzidis A. (2018) Emerging concepts in non-invasive monitoring of Crohn's disease. *Therapeutic Advances in Gastroenterology*, 11: 1756284818769076. doi: 10.1177/1756284818769076.

Mayer E.A., Knight R., Mazmanian S.K., Cryan J.F., Tillisch K. (2014) Gut microbes and the brain: Paradigm shift in neuroscience. *The Journal of Neuroscience*, 34: 15490–15496.

Menni C., Jackson M.A., Pallister T., Steves C.J., Spector T.D., Valdes A.M. (2017) Gut microbiome diversity and high-fibre intake are related to lower long-term weight gain. *International Journal of Obesity*, 41: 1099–1105. doi: 10.1038/ijo.2017.66.

Mesa M.D., Loureiro B., Iglesia I., Fernandez Gonzalez S., Llurba Olivé E., García Algar O., Solana M.J., Cabero Perez M.J., Sainz T., Martinez L., Escuder-Vieco D., Parra-Llorca A., Sánchez-Campillo M., Rodriguez Martinez G., Gómez Roig D., Perez Gruz M., Andreu-Fernández V., Clotet J., Sailer S., Iglesias-Platas I., López-Herce J., Aras R., Pallás-Alonso C., de Pipaon M.S., Vento M., Gormaz M., Larqué Daza E., Calvo C., Cabañas F.(2020) The Evolving Microbiome From Pregnancy To Early Infancy: A Comprehensive Review. *Nutrients*, 12(1): 133. doi: 10.3390/nu12010133.

Michael O.S., Dibia C.L., Soetan O.A., Adeyanju O.A., Oyewole A.L., Badmus O.O., Adetunji C.O., Soladoye A.O. (2020) Sodium acetate prevents nicotine-induced cardiorenal dysmetabolism through uric acid/creatine kinase-dependent pathway. *Life Sciences*, 257: 118127. doi: 10.1016/j.lfs.2020.118127.

Miller W.B. (2016) The eukaryotic microbiome: Origins and implications for fetal and neonatal life. *Frontiers in Pediatrics*, 4: 96. doi: 10.3389/fped.2016.00096.

Miro-Blanch J., Yanes O. (2019) Epigenetic regulation at the interplay between gut microbiota and host metabolism. *Frontiers in Genetics*, 10: 638.

Moloney R.D., Desbonnet L., Clarke G., Dinan T.G., Cryan J.F. (2014) The microbiome: stress, health and disease. *Mammalian Genome*, 25(1–2): 49–74.

Montagne A., Barnes S.R., Sweeney M.D., Halliday M.R., Sagare A.P., Zhao Z., Toga A.W., Jacobs R.E., Liu C.Y., Amezcua L., Harrington M.G., Chui H.C., Law M., Zlokovic B.V. (2015) Blood-brain barrier breakdown in the aging human hippocampus. Neuron, 85: 296–302.

Nayfach S., Rodriguez-Mueller B., Garud N., Pollard K.S. (2016) An integrated metagenomics pipeline for strain profiling reveals novel patterns of bacterial transmission and biogeography. *Genome Research*, 26: 1612–1625.

Nicholson J.K., Holmes E., Kinross J., Burcelin R., Gibson G., Jia W., Pettersson S. (2012) Host-gut microbiota metabolic interactions. *Science*, 336: 1262–1267.

Olaniyi K.S., Amusa O.A. (2020) Sodium acetate-mediated inhibition of histone deacetylase alleviates hepatic lipid dysregulation and its accompanied injury in streptozotocin-nicotinamide-induced diabetic rats. *Biomedicine and Pharmacotherapy*, 128: 110226.

Olaniyi K.S., Amusa O.A., Areola E.D., Olatunji L.A. (2020) Suppression of HDAC by sodium acetate rectifies cardiac metabolic disturbance in streptozotocin–nicotinamide-induced diabetic rats. *Experimental Biology and Medicine (Maywood)*, 245(7): 667–676. doi: 10.1177/1535370220913847.

Pacheco A.R., Barile D., Underwood M.A., Mills, D.A. (2015) The impact of the milk glycobiome on the neonate gut microbiota. *Annual Review of Animal Biosciences*, 3: 419–445. doi: 10.1146/annurev-animal-022114-111112.

Parent A., Carpenter M.B. (1996) *Carpenter's Human Neuroanatomy*. Baltimore, MD: Williams & Wilkins.

Parlee S.D., MacDougald O.A. (2014) Maternal nutrition and risk of obesity in offspring: The Trojan horse of developmental plasticity. *Biochimica et Biophysica Acta*, 1842: 495–506. doi: 10.1016/j.bbadis.2013.07.007.

Perez P.F., Doré J., Leclerc M., Levenez F., Benyacoub J., Serrant P., Segura-Roggero I., Schiffrin E.J., Donnet-Hughes A. (2007) Bacterial imprinting of the neonatal immune system: Lessons from maternal cells? *Pediatrics*, 119(3): e724–e732. doi: 10.1542/peds.2006-1649.

Perry R.J., Peng L., Barry N.A., Cline G.W., Zhang D., Cardone R.L., Petersen K.F., Kibbey R.G., Goodman A.L., Shulman G.I. (2016) Acetate mediates a microbiome-brain-b-cell axis to promote metabolic syndrome. *Nature*, 534: 213–217.

Poirier L.A., Wise C.K., Delongchamp R.R., Sinha R. (2001) Blood determinations of S -adenosylmethionine, S -adenosylhomocysteine, and homocysteine: Correlations with diet. *Cancer Epidemiology, Biomarkers and Prevention*, 10: 649–655.

Portha B., Fournier A., Kioon M.D., Mezger V., Movassat J. (2014) Early environmental factors, alteration of epigenetic marks and metabolic disease susceptibility. *Biochimie*, 97: 1–15. doi: 10.1016/j.biochi.2013.10.003.

Qin Y., Roberts J.D., Grimm S.A., Lih F.B., Deterding L.J., Li R., Chrysovergis K., Wade P.A. (2018) An obesity-associated gut microbiome reprograms the intestinal epigenome and leads to altered colonic gene expression. *Genome Biology*, 19(1): 7. doi: 10.1186/s13059-018-1389-1.

Qin Y., Wade P.A. (2018) Crosstalk between the microbiome and epigenome: Messages from bugs. *Journal of Biochemistry*, 163(2): 105–112. doi: 10.1093/jb/mvx080.

Robertson R.C., Manges A.R., Finlay B.B., Prendergast, A.J. (2019) The human microbiome and child growth – first 1000 days and beyond. *Trends in Microbiology*, 27(2): 131–147. doi: 10.1016/j.tim.2018.09.008.

Romano-Keeler J., Weitkamp J.H. (2015) Maternal influences on fetal microbial colonization and immune development. *Pediatric Research*, 77(1–2): 189–195.

Russell W.R., Gratz S.W., Duncan S.H., et al. (2011) High-protein, reduced-carbohydrate weight-loss diets promote metabolite profiles likely to be detrimental to colonic health. *American Journal of Clinical Nutrition*, 93(5): 1062–1072. doi: 10.3945/ajcn.110.002188.

Sampson T.R., Mazmanian S.K. (2015) Control of brain development, function, and behavior by the microbiome. *Cell Host and Microbe*, 17: 565–576.

Sharma M., Li Y., Stoll M.L., Tollefsbol T.O. (2019) The epigenetic connection between the gut microbiome in obesity and diabetes. *Frontiers in Genetics*, 10: 1329. doi: 10.3389/fgene.2019.01329.

Smith P.A. (2015) The tantalizing links between gut microbes and the brain. *Nature*, 526: 312–314.

Smith C.L., Dickinson P., Forster T., Craigon M., Ross A., Khondoker M.R., France R., Ivens A., Lynn D.J., Orme J., Jackson A., Lacaze P., Flanagan K.L., Stenson B.J., Ghazal P. (2015) Identification of a human neonatal immune metabolic network associated with bacterial infection. *Nature Communications*, 5: 4649.

Stiemsma L.T., Michels K.B. (2018) The role of the microbiome in the developmental origins of health and disease. *Pediatrics*, 141: e20172437.

Stilling R.M., Dinan T.G., Cryan J.F. (2014) Microbial genes, brain & behaviour - epigenetic regulation of the gut-brain axis. *Genes, Brain, and Behavior*, 13: 69–86.

Tilg H., Zmora N., Adolph T.E., Elinav E. (2020) The intestinal microbiota fuelling metabolic inflammation. *Nature Reviews Immunology*, 20(1): 40–54. doi: 10.1038/s41577-019-0198-4.

Turnbaugh P.J., Ridaura V.K., Faith J.J., Rey F.E., Knight R., Gordon J.I. (2009) The effect of diet on the human gut microbiome: A metagenomic analysis in humanized gnotobiotic mice. *Science Translational Medicine*, 1(6): 6ra14. doi: 10.1126/scitranslmed.3000322.

Ursell L.K., Metcalf J.L., Parfrey L.W., Knight R. (2012) Defining the human microbiome. *Nutrition Reviews*, 70(Suppl 1): S38–S44. doi: 10.1111/j.1753-4887.2012.00493.x.

Van Winckel M., De Bruyne R., Van De Velde S., Van Biervliet S. (2009) Vitamin K, an update for the paediatrician. *European Journal of Pediatrics*, 168(2): 127–134. doi: 10.1007/s00431-008-0856-1.

Virtue A.T., McCright S.J., Wright J.M., et al. (2019) The gut microbiota regulates white adipose tissue inflammation and obesity via a family of microRNAs. *Science Translational Medicine*, 11(496): eaav1892. doi: 10.1126/scitranslmed.aav1892.

Wang Q., Wang K., Wu W., Giannoulatou E., Ho J.W.K., Li L. (2019) Host and microbiome multiomics integration: Applications and methodologies. *Biophysical Reviews*, 11(1): 55–65. doi: 10.1007/s12551-018-0491-7.

Weaver I.C., Korgan A.C., Lee K., Wheeler R.V., Hundert A.S., Goguen D. (2107) Stress and the emerging roles of chromatin remodeling in signal integration and stable transmission of reversible phenotypes. *Frontiers in Behavioral Neuroscience*. 11: 41.

Wen X., Hu X., Miao L., Ge X., Deng Y., Bible P.W., Wei L. (2018) Epigenetics, microbiota, and intraocular inflammation: New paradigms of immune regulation in the eye. *Progress in Eye and Retinal Research*, 64: 84–95. doi: 10.1016/j.preteyeres.2018.01.001.

Winek K., Dirnagl U., Meisel A. (2016) The gut microbiome as therapeutic target in central nervous system diseases: Implications for stroke. *Neurotherapeutics: The Journal of the American Society for Experimental NeuroTherapeutics*, 13(4): 762–774.

Yang C.S., Lambert J.D., Sang S. (2009) Antioxidative and anti-carcinogenic activities of tea polyphenols. *Archives of Toxicology*, 83(1): 11–21.

Yatsunenko T., Rey F.E., Manary M.J., et al. (2012) Human gut microbiome viewed across age and geography. *Nature*, 486(7402): 222–227. doi: 10.1038/nature11053.

Yu R., Wang X., Moazed D. (2018) Epigenetic inheritance mediated by coupling of RNAi and histone H3K9 methylation. *Nature*, 558(7711): 615–619.

Zhang Z., Tang H., Chen P., Xie H., Tao Y. (2019) Demystifying the manipulation of host immunity, metabolism, and extraintestinal tumors by the gut microbiome. *Signal Transduction and Targeted Therapy*, 4(1): 41. doi: 10.1038/s41392-019-0074-5.

Zhu S., Jiang Y., Xu K., Cui M., Ye W., Zhao G., Jin L., Chen X.J. (2020) The progress of gut microbiome research related to brain disorders. *Neuroinflammation*, 17(1): 25. doi: 10.1186/s12974-020-1705-z.

7 Probiotic Research Studies That Led to the Discovery of *Lactobacillus pentosus* KCA1

Kingsley C. Anukam
Nnamdi Azikiwe University
Uzobiogene Genomics

CONTENTS

7.1 Background to Probiotic Research in Nigeria 121
7.2 Preliminary Probiotics Research Studies in Nigeria 123
7.3 Studies on Safety of Probiotics 125
7.4 Determining Whether the Vagina of African/Nigerian Women Are Colonized by Lactobacilli? 126
7.5 Human Clinical Studies with Probiotics 127
7.6 Probiotic Properties of *Lactobacillus* species Isolated from the Vagina of Healthy Nigerian Women 128
7.7 Growth Kinetics of *Lactobacillus plantarum* KCA1 in Vaginal Defined Medium with Peptone (VDMP) 129
7.8 The Challenges Encountered in *Lactobacillus plantarum* KCA1 Green Fluorescence Protein (gfp) Electroporation 131
7.9 Emergence of and Sequencing of the First *Lactobacillus pentosus* KCA1 from Nigeria 132
7.10 Conclusion 133
References 133

7.1 BACKGROUND TO PROBIOTIC RESEARCH IN NIGERIA

In the late 1990s, when Human Immunodeficiency Virus (HIV)/Acquired Immunodeficiency Syndrome (AIDS) was ravaging African continent, I was lucky to have come across a pamphlet from the United Nations. The pamphlet contained information on the possible preventive measures that could be utilized to reduce the risk and the burden of HIV/AIDS. One of the preventive measures was the use of microbicides that could be applied intravaginally to inactive HIV from gaining access to vaginal cells. Most scientists were aware of the microbicide, Nonoxynol-9 which was touted as a possible solution, but it does not adequately address intervention

DOI: 10.1201/9781003180241-7

approaches, instead it exacerbated the risk of acquisition of HIV/AIDS. As I read through the pamphlet, I became more curious when I came across lactobacilli being suggested as a possible microbicide that could be used to fight HIV/AIDS. I became more inclined to retrieve any useful information on how lactobacilli could be used to prevent infection, especially in the urogenital tracks of women. After my Master of Science (MSc) graduation in 1997, from the Department of Pharmaceutical Microbiology, University of Benin, with special interest in Antimicrobial Agents and Chemotherapy, I had "gut" feeling of going extra mile to contribute research knowledge that could make a difference in health care. In a quest for current research, I became a full member of the American Society of Microbiology (ASM) in 1999 after payment of US$100.00 and I subscribed to receive monthly, the *Journal of Antimicrobial Agents and Chemotherapy* and *Journal of Clinical Microbiology.* I remember a close friend retorted "Why are you wasting your money on all these international journals"; I was admitted for a PhD program in 2000 at the Department of Pharmaceutical Microbiology, Faculty of Pharmacy, University of Benin, Nigeria. Initially my supervisor suggested a research project on pertussis and parapertussis in previously vaccinated adolescents and association with HIV infection. I started the project with zeal to the extent that preliminary finding was published in *World Journal of Microbiology and Biotechnology* (Anukam et al. 2004a). However, I could not go further with pertussis research due to persistent lack of basic infrastructures in the faculty/departmental research laboratories in the universities across the breadth of Nigeria. I approached my supervisor Professor Emmanuel Osazuwa that I could not continue with the research topic he suggested. He allowed me to come up with any research topic that I could do, and he sternly warned me that I should be ready to be the "driver" on any topic of my choice while he stays at the back seat to provide professional/academic advise if need be. The approval of my supervisor to develop any other research topic after over 2 years of research on pertussis challenged my efforts head-on to get more information on lactobacilli as a potential microbicide. Information on any research topic could be retrieved in the library, but regrettably, the libraries in Nigerian universities are very old, to the extent that I could only get very little or no useful information on the use of lactobacilli as a microbicide. The only option available then was to turn to the internet, but again, there was very limited internet availability in the University. Only the Vice-Chancellor's office has access to the internet. My other option was to source for commercial internet providers in Benin City. The only one available was in Prest-Motel along Airport Road in Benin City. I was made to register with One thousand Five Hundred Naira (N1,500.00) in order to be allowed to obtain yahoo email address, which is supposed to be free and to browse the internet for 1 hour. I was determined as I paid to get a yahoo email (anukamkc@yahoo.com, which I still use till this day) and browsed for 1 hour looking for information on lactobacilli . As providence will have it, I downloaded some published articles on lactobacilli and saved them in my diskette (Memory stick or Flash drive was not available/popular as at that time between 2000 and 2001). I had a desktop computer (Pentium III) as at 1998, which I used for private research work at Chidak Medical Diagnostic Laboratories, Benin City. So, opening the diskette was not a problem to me, coupled with the fact that I had a desk jet printer. I was able to print out some articles on lactobacilli. One of the published articles that made me to ponder for several months was from Antonio et al. (1999). The authors used culture methods, followed by whole-chromosomal DNA probes to determine the types of lactobacilli that colonize the vagina of healthy American women, which included, Whites, Asians, Hispanics and Black women. The study concluded that colonization was primarily by *Lactobacillus crispatus* or *Lactobacillus jensenii* and was associated with being a white, and that Black women's vagina is less colonized by lactobacilli. Other studies that followed subsequently cited the article as the reason black women are at higher risk of bacterial vaginosis (BV) and HIV/AIDS (Cohn and Clark 2003). This publication was a "game-changer" for me and formed the basis of my PhD research as I set out to determine if truly Black women's vagina are devoid of colonization by lactobacilli. Another article I read that raised my curiosity reported that displacement of vaginal lactobacilli results in developing BV, which is a major risk for acquisition of sexually transmitted diseases and HIV/AIDS (Schwebke 2001).

In 2001, I read a publication from the Food and Agriculture Organization (FAO) of the United Nations and the World Health Organization (WHO) highlighting the health benefits of lactic acid bacteria that defined probiotics as "Live microorganisms which when administered in adequate amounts confer a health benefit on the host" (FAO/WHO 2001). The word "probiotics" was very strange, and on reading further, it became clearer that lactobacilli and other lactic acid producing bacteria could be used as probiotics. In a paper I authored in 2007, historical journey of the word "probiotics" and the use of lactic acid bacteria was traced back to Elie Metchnikoff, who was presented with a Nobel Prize for medicine in 1908 (Anukam and Reid 2007).

7.2 PRELIMINARY PROBIOTICS RESEARCH STUDIES IN NIGERIA

In the process of seeking for more information on lactobacilli now referred to in this article as probiotics, Dr. Gregor Reid, who chaired the FAO/WHO Expert Consultation in 2001, came on focus through the internet as a result of his several earlier publications on Probiotics. I contacted him in 2001. With persistence for more information to the point of "hormonal fluctuations" he was kind enough to send some of his research works on lactobacilli to me.

As a prelude, I made a presentation titled "Can probiotics replace antibiotics in the future?" at the 39th Annual Scientific Conference of the Association of Medical Laboratory Scientists of Nigeria. The whole audience were kept in rapt attention during the presentation. The response from the audience revealed that over and above 90% have not heard of the word "probiotics", but they are aware of lactobacilli as a normal microbiota, not associated with infections. Questions on safety were asked and possible translocation to the visceral organs. I put it to them that "it is the test of fire that makes fine steel." For a good probiotic product, safety profiles must be established.

With the exciting interest shown by some scientists during the conference, I published a comprehensive review on "Probiotic functions of lactobacilli" in the *Journal of Medical Laboratory Science* (Anukam 2004). The summary of the review is hereby presented below:

It should be noted that a strain of probiotic microorganism can exhibit different characteristics in order to survive or strive with other microbes in a particular niche.

1. Adherence factors

   Several studies have reported the capability of lactobacilli to adhere to uroepithelial cells, but the adherence capacity was found to vary among strains. For instance, lactobacilli may utilize several mechanisms to attach to surfaces with capsular and fimbriae structures.
2. Competitive exclusion and coaggregation

   The capability of lactobacilli to inhibit the adhesion of uropathogens by competitive exclusion has been reported several in vitro and in vivo studies. The ability to coaggregate with uropathogens might be important in maintaining an ecological balance within the urogenital tract. This, together with the production of inhibitory substances, might constitute an important host defense mechanism.
3. Biosurfactant production

   Some lactobacilli such as *Lactobacillus pentosus* KCA1 produce biosurfactants that inhibit the adhesion of uropathogens to surfaces. The anti-adhesive properties of biosurfactants produced by some *Lactobacillus* strains are applied in human sites where pathogenic microbes adhere and colonize.
4. Production of lactic acids

   Lactic acids produced by lactobacilli are responsible for the maintenance of low pH in the vagina of premenopausal women, which is understood to be one important mechanism controlling vaginal colonization by uropathogens.
5. Hydrogen peroxide

   Hydrogen peroxide is produced by some, but not all strains of lactobacilli are thought to play significant role in inhibiting the growth of urogenital pathogens. In vitro, hydrogen peroxide has been shown to kill various bacteria such as *Gardnerella vaginalis*, *Neisseria gonorrhoeae* and *Prevotella bivia*.
6. Bacteriocins

   Besides lactic acid and hydrogen peroxide, lactobacilli also produce a number of bacteriocins. Bacteriocin compounds from several lactobacilli strains were found to have an inhibitory activity against several bacterial species such as *Pseudomonas* spp., *Staphylococcus* spp., *Streptococcus* spp., and *Enterobacteriaceae*.
7. Acid tolerance

   The presence of lactobacilli has previously been shown to correlate with a low vaginal pH. Acidity has long been thought to be one of the protective mechanisms of the vagina, and probiotic microbes usually possess tolerance to high acidity.

I decided to investigate the level of knowledge on probiotics at the University of Benin in 2004. A survey questionnaire was designed to test the local knowledge on the idea of probiotics. The objective of the study was to test their knowledge on Probiotics and their willingness to use such products should they be introduced into the country as a healthcare product. The survey enrolled 280 female university

TABLE 7.1
Safety and Other Issues Raised about Probiotic Use by 25% of the Questions Raised by Respondents

- Would vaginal probiotics be safe for all classes of women?
- Can I insert probiotic capsules inside my vagina during the menstrual period and what will be the effect?
- Would vaginal probiotics be a form of vaginal cleansing or douching?
- Is the probiotic useful for healthy vagina?
- How can you determine the quantity of lactic acid bacteria in my vagina before you recommend the probiotics for me?
- Would it not lead to excessive bacteria in my vagina?
- How can I insert another person's lactobacilli in my body?
- What is the long-term effect of continuous ingestion or instillation of vaginal probiotics and even in yogurts?
- How would my husband react if he noticed that I inserted lactic acid bacteria in my vagina?
- Would it increase vaginal lubrication or lead to vaginal dryness?

Anukam et al. (2004); *Sex. Trans. Dis.*; 31(8): 460–464.

students aged 26–30 years, 55% said they were at risk of HIV. The survey showed that over 80% of the women would welcome probiotics in yogurt form and/or oral form to improve their vaginal health. All the students (100%) were not familiar with the word "probiotics". Most would buy probiotics as yogurt (40%) ahead of oral or vaginal caps. Over 36% were willing to use probiotics as a daily self-care with 64% expressing desire to purchase the probiotics in pharmacies; 19% supermarkets; and 15% in clinics. Interestingly, 39.6% of the female respondents were willing to pay between $0.08 and $0.75 dollars for probiotic products. Over 25% raised some concerns about the safety of probiotics. Some asked "Would it not lead to excessive bacteria in my vagina or stomach" as represented in Table 7.1.

The result of the survey was published in *Sexually Transmitted Diseases* (Anukam et al. 2004).

Public records have shown that a large number of published data on probiotics exist in developed countries especially in Europe and North America. While some concerns were raised by female students in our previous survey study, we decided to measure the level of knowledge on probiotics from Nigerian clinicians (Anukam et al. 2004). The study revealed that over 95% of Nigerian clinicians surveyed are not familiar with the word "probiotics", but they all expressed similar enthusiasm like premenopausal female students (Anukam et al. 2004a).

7.3 STUDIES ON SAFETY OF PROBIOTICS

In order to establish safety of probiotics, due to some concern raised in our previous survey studies, we hypothesized in an animal model that daily feeding of clinically tested probiotics will not significantly alter the hematological parameters. The Sprague Dolly albino rats were administered orally daily dose (10^9 cfu/mL) of

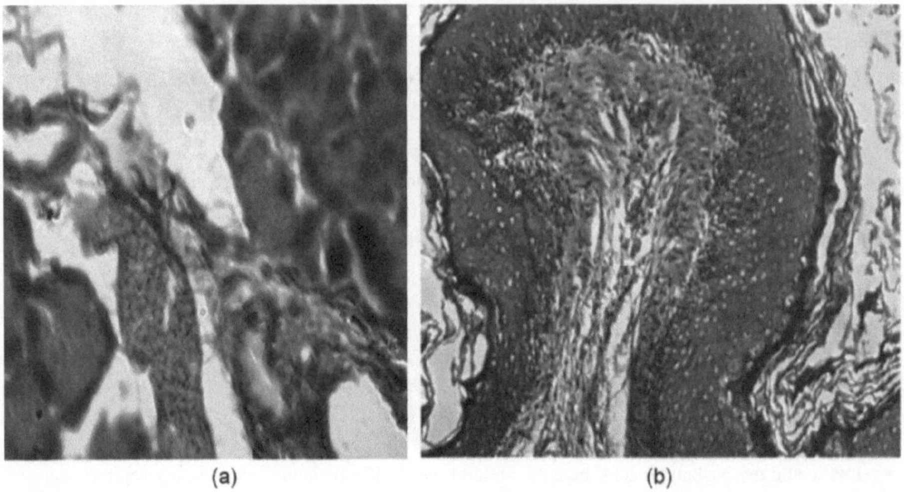

FIGURE 7.1 (a) Indomethacin-induced gastric erosion. (b) Restoration of IBD after feeding with GR-1/RC-14.

the probiotic strains using oro-gastric tube for 21 days. Blood samples were collected into potassium EDTA containers and 12 parameters were determined with auto-hematological analyzer. Results showed no significant difference between rats fed with probiotics when compared with the controls, thus predicating that daily feeding of probiotics GR-1 and RC-1 may not modify the peripheral red blood cells. Thus, dousing concerns with regard to inflammatory response to probiotics (Anukam et al. 2004). In another unpublished study, we induced inflammatory bowel disease (IBD) with Indomethacin in rats showing gastric erosion, and thereafter they were treated/fed with GR-1/RC-14 for 7 days (Figure 7.1a and b).

In another study, we hypothesized that improvement of appetite and health of newborns could be achieved by use of probiotics as a food supplement. This was based on the fact that malnutrition and infection of the vagina have been found to be responsible for poor quality of life as a result of preterm birth of newborns and low birth weight of infants. The study revealed that rats treated with probiotics had a significant improvement in appetite when compared with controls ($P = 0.006$). In addition, the birth weight of new born rats whose mothers were fed with probiotics increased significantly ($P = 0.01$) (Anukam et al. 2005).

7.4 DETERMINING WHETHER THE VAGINA OF AFRICAN/NIGERIAN WOMEN ARE COLONIZED BY LACTOBACILLI?

Having established the safety of lactobacilli, though in rats, my next focus was to determine whether African women's vaginas are colonized by lactobacilli. I ordered the growth media (MRS-deManRogosa Sharpe) agar and broth for cultivating lactobacilli directly from England. Initially, I had problems growing lactobacilli from the High Vaginal Swabs of healthy women, until I found that Lactobacilli grow better in micro aerophilic environment. I improvised the micro environment with candle jar and incubated the culture plates at 37°C for 24–28 hours. The colonies that grew were gram stained and catalase test done for presumptive identification of lactobacilli.

Scientific studies in Africa especially Nigeria are hampered by many factors, which include but not limited to lack of scientific research infrastructures, erratic power supply and extreme poor funding for research. Most Universities in Nigeria do not have basic research facilities such as Polymerase Chain Reaction (PCR), Denaturing Gradient Gel Electrophoresis (DGGE), amplicon or full DNA sequencing facilities. Identifying those presumptive lactobacilli from apparently healthy women which I grew in MRS agar and broth became another herculean task that must be tackled.

Establishing the real Identity of lactobacilli present in Africa most importantly Nigerian women was my primary objective and to provide scientific answer to the question whether African women harbor lactobacilli in their vagina contrary to previous published report from America that Africa-American women do not harbor lactobacilli in their vagina unlike white women. Can lactobacilli from Africa, especially Nigerian women that constitute 25% of the continent's population, exhibit properties with probiotic properties? In search for answers to the above-mentioned questions, by serendipity of courage, personal sacrifice and persistence, I was offered a PhD student scholarship in 2004 by Dr. Gregor Reid at the Lawson Health Research Institute/University of Western Ontario to enable me analyze the vaginal samples.

The study determined whether *Lactobacillus* species that colonized the vagina of Nigerian women differ from species that colonized Caucasian women as described in previous studies. In the study (Anukam et al. 2006c), vaginal samples were collected from 241 healthy premenopausal women and bacterial DNA from the samples were sequenced. Results revealed that most Nigerian women sampled were colonized by *Lactobacillus* species, similar to previous findings in Canadian and Swedish women (Burton et al. 2003; Falsen et al. 1999; Vasquez et al. 2002). The findings from the study showed that irrespective of geographic, racial, social and economic differences, the major lactobacilli colonizing the vagina of Nigerian women is similar to *Lactobacillus* species colonizing the vagina of women from the Western world. First objective has been accomplished. The seminar/defense of my PhD thesis in 2005 was indeed very interesting as it turned out to be a teaching class for both students and academic staff of the Department of Pharmaceutical Microbiology, University of Benin. I was truly the "driver" as predicted by my supervisor.

7.5 HUMAN CLINICAL STUDIES WITH PROBIOTICS

In order to begin to understand if the use probiotics could have a place in prevention of HIV/AIDS, three studies were undertaken. The first was a randomized, placebo-controlled trial using probiotics with antibiotics to cure BV, a major risk factor for HIV in women (Sha et al. 2005). In the study, we enrolled 125 women with BV. The subjects were given a single oral dose of metronidazole, and oral *L. rhamnosus* and *L. reuteri*. Primary outcome was cure of BV as 84.8% of the women had no symptoms at day 30. Higher proportion of *Lactobacillus* species were recovered from the vagina of 96% probiotic-treated subjects compared to 53% controls at day 30 (Figure 7.2). The study demonstrated effective use of lactobacilli in reducing

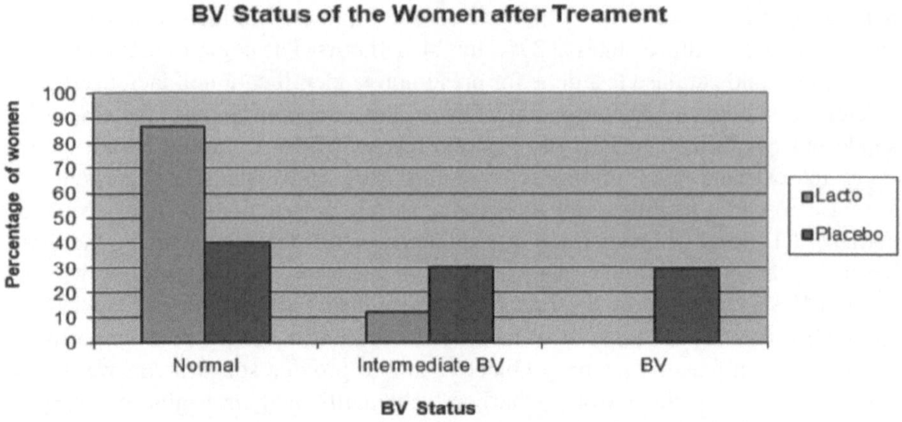

FIGURE 7.2 Showing BV status of women treated with probiotics and with placebo. (Anukam et al. (2006); *Microbes Infection*; 8(6):1450–1454.)

BV in Nigerian women (Anukam et al. 2006a). This restoration of a normal microbiota may have an important impact in efforts to diminish the risk associated with HIV acquisition in sub-Saharan African, especially Nigerian women.

Given that some strains of probiotic lactobacilli were able to colonize the vagina and inhibit urogenital pathogens, we tested whether probiotic lactobacilli can be applied in the treatment of symptomatic BV. The second study was to determine whether the use of intravaginal application of *Lactobacillus* species could cure BV, compared to standard metronidazole gel. The study enrolled 20 women with BV, and they were treated for 5 days with vaginal *Lactobacillus* GR-1/RC-14 capsules (2×10^9/day). Improvement of BV status was observed in the subjects at day 7 and 30 showing association between clinical urogenital health, Nugent score and negative sialidase test (Anukam et al. 2006b).

The third study was undertaken on 24 HIV/AIDS subjects to assess safety and tolerance of probiotics similar to previous studies (Wolf et al. 1998). We decided to use yogurt fortified with *Lactobacillus* strains daily for 15 days in HIV/AIDS patients with moderate diarrhea and CD4 cell counts between 206 and 520 cells/μL. The CD4 cell count increased in 11/12 probiotic-treated subjects compared to 3/12 in control group (Anukam et al. 2008). The study exemplified the need to bring clinically tested probiotics to Nigeria.

7.6 PROBIOTIC PROPERTIES OF *LACTOBACILLUS* SPECIES ISOLATED FROM THE VAGINA OF HEALTHY NIGERIAN WOMEN

The concept of probiotics had gained an increasing momentum in the past four decades in the Western hemisphere, yet the idea is not on the radar screen of most African countries, especially Nigeria that is in great need of probiotic benefits. From the previous study (Anukam et al. 2005), *Lactobacillus* species that grew in MRS culture

TABLE 7.2
Summary of Biosurfactant and H_2O_2 Production by Isolated *Lactobacillus* Species

Lactobacillus Species	Biosurfactant Producer	H_2O_2 Producer
Lactobacillus plantarum	Positive	Positive[a]
Lactobacillus fermentum	Positive	Positive[a]
Lactobacillus gasseri	Negative	Positive
Lactobacillus rhamnosus	Negative	Negative
Lactobacillus crispatus	Negative	Negative
Lactobacillus vaginalis	Negative	Negative
Lactobacillus rhamnosus GR-1	Positive	Positive

[a] *Lactobacillus plantarum* and *L. fermentum* were selected for further probiotic properties.

medium were identified with Lactobacillus primers using PCR and sequencing of the 16S rRNA with 37830xl ABI sequencer. Two *Lactobacillus* species (*Lactobacillus plantarum* and *Lactobacillus fermentum*) were tested for probiotic properties as shown in Table 7.2 (Anukam and Koyama 2007; Anukam and Reid 2007a).

BV prevalence has been reported between 29% and 50%, and current antibiotic treatment does not cure the condition, nor prevent recurrence (Bradshaw et al. 2006). Our clinical study showing that probiotic lactobacilli can augment antibiotic efficacy was investigated in an in vitro study (Anukam and Reid 2008).

7.7 GROWTH KINETICS OF *LACTOBACILLUS PLANTARUM* KCA1 IN VAGINAL DEFINED MEDIUM WITH PEPTONE (VDMP)

We determined the growth curves of *L. plantarum* KCA1 and other strains by microdilution method in VDMP (Figure 7.3a). Growth kinetics of *Lactobacillus plantarum* KCA1 in VDMP was very high in comparison with *G. vaginalis*, thus suggesting that *Lactobacillus plantarum* KCA1 has the potential to outcompete *G. vaginalis* that causes BV in women. *Gardnerella vaginalis* ATCC 14018, an organism found to be associated with BV, and *L rhamnosus* GR-1 as well as clinical isolate of *Lactobacillus plantarum* KCA1 were inoculated into microplates containing different concentrations of metronidazole. Growth of the different strains was distinctly inhibited at high antibiotic concentrations, with a significant ($P = 0.005$) stimulatory effect observed for *Lactobacillus plantarum* KCA1 at 256 μg/mL ($P = 0.02$). At concentrations ≥ 4,000 μg/mL, there was marked growth inhibition of more than 71% on the two lactobacilli strains, while *Gardnerella vaginalis* was inhibited over 70% at ≥3,000 μg/mL (Figure 7.3b and c). The finding that metronidazole concentrations greater than 1,000 μg/mL inhibited *Gardnerella vaginalis* and two lactobacilli strains suggested that the dose of 500 mg orally or 0.75% (7.5 mg/g) intravaginally used to treat BV can also kill both pathogens and normal microbiota. The use of probiotic lactobacilli, or lactic acid gel, or low-dose antibiotic need to be considered to better manage this highly prevalent condition and reduce the risk of its complications.

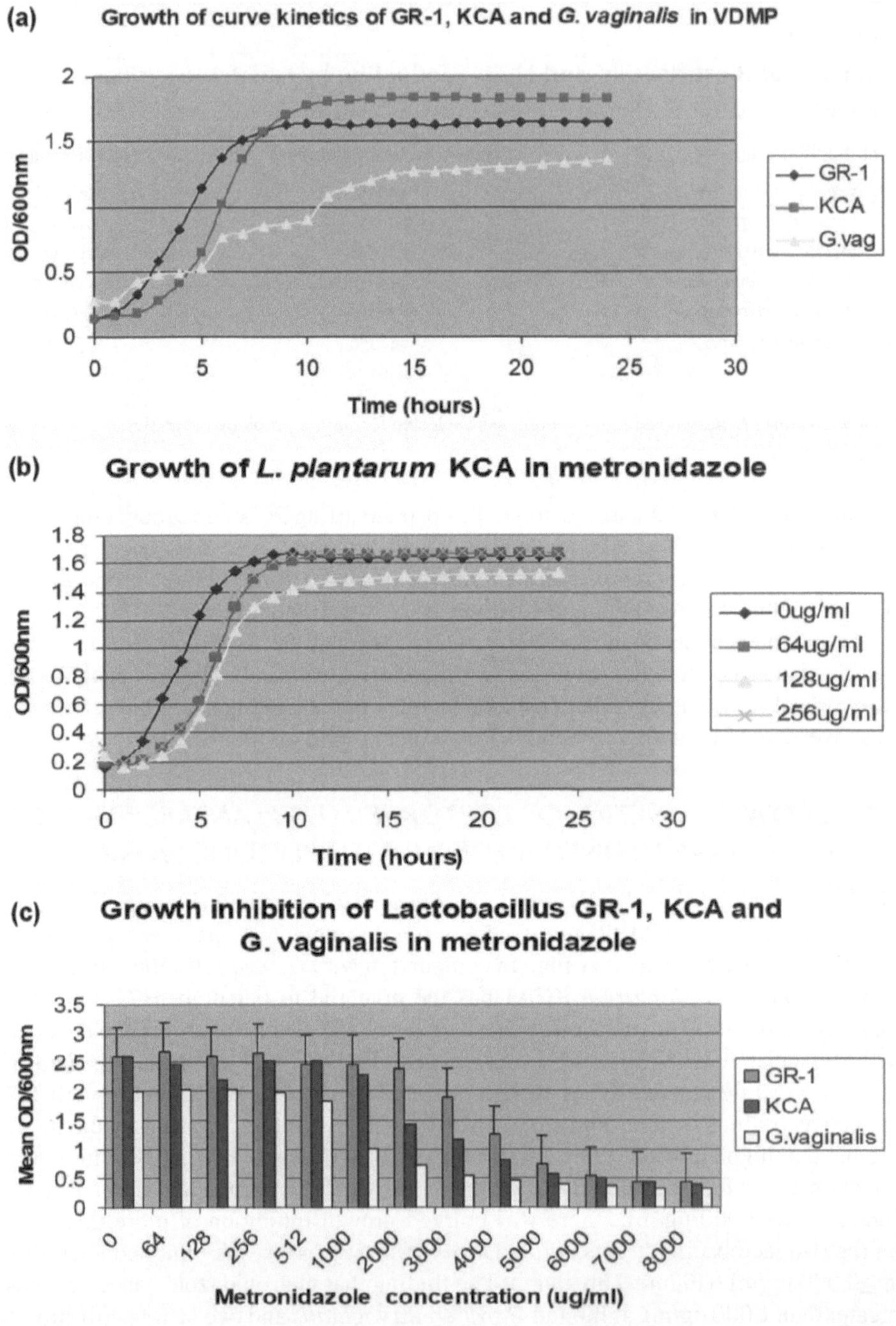

FIGURE 7.3 (a) Showing growth kinetics of *L. plantarum* KCA1, GR-1 and *G. vaginalis*. (b) Showing growth of *L. plantarum* KCA1 in various concentrations of metronidazole. (c) Inhibition of *Lactobacillus* species and *G vaginalis*. in metronidazole. Anukam KC, Reid G (2008); *Microbial Ecology in Health and Disease*; 20(1):48–52.

7.8 THE CHALLENGES ENCOUNTERED IN *LACTOBACILLUS PLANTARUM* KCA1 GREEN FLUORESCENCE PROTEIN (GFP) ELECTROPORATION

During my post-doctoral research fellowship at the Lawson Health Research Institute/University of Western Ontario, between 2006 and 2008, one of the challenges I encountered was how to determine if a gfp can be electroporated into *Lactobacillus rhamnosus* GR-1. This was based on the fact that *Lactobacillus rhamnosus* GR-1 has been detected in the vagina after oral administration. Some studies have shown that gfp expressing lactobacilli can be used to trace movement of bacteria in the gut. However, no studies have yet examined if lactobacilli can ascend from the rectum to the vagina. In order to test this hypothesis, we decided to determine if a gfp can be electroporated into *L. rhamnosus* GR-1 strain. We designed the study to express the gfp gene under the control of the nisin promoter in *L. rhamnosus* GR-1 and in *L. plantarum* KCA1. The study involved growing overnight cultures of *L. rhamnosus* GR-1 and *L. plantarum* KCA1, and the two strains were electroporated with gfp plasmid pSMJJ8. The growth of the gfp-GR-1 and gfp-L_plantarum-KCA1 were monitored by measuring the absorbance with a Multiskan Ascent V1. The *Lactobacillus* strains (1 mL) were washed in phosphate buffered saline (PBS) and the cells were suspended in PBS for fluorescence microscopy examination.

GR-1-gfp positive cells grown aerobically were observed. There were no *L. plantarum* KCA1-gfp positive cells (Figure 7.4a and b). Growth of the GR-1-gfp under the promoter nisin induction was influenced by the concentration of the nisin. Decreased growth and fluorescence were observed at higher concentrations of the bacteriocin. GR-1-gfp and KCA1-gfp cultures grown under anaerobic condition did not show any sign of fluorescence. The reason for no positive *L. plantarum* KCA1-gfp could not

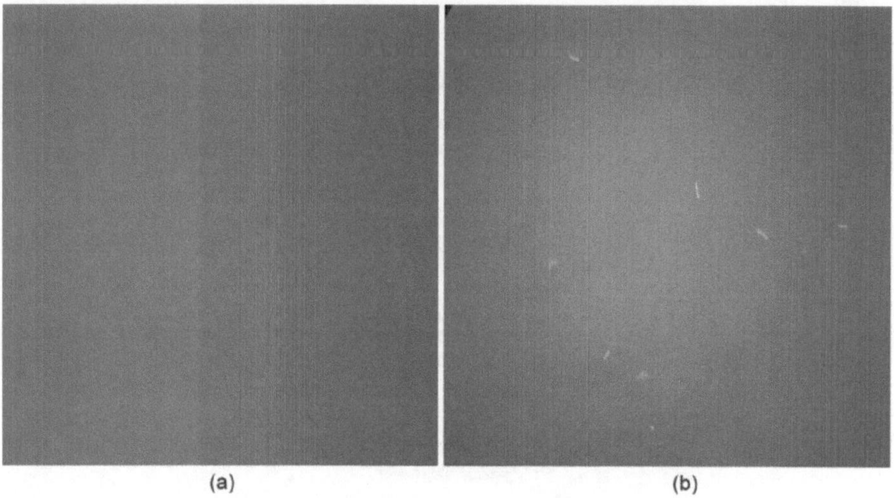

FIGURE 7.4 (a) Gfp-*L. plantarum* KCA1 negative cells. (b) Gft-*L. rhamnosus* GR-1 positive cells.

be established after several repeated electroporation with the gfp plasmid pSMJJ8. I decided to store *L. plantarum* KCA1 and other strains at −80°C facility at the Lawson Research Institute for future investigations.

7.9 EMERGENCE OF AND SEQUENCING OF THE FIRST *LACTOBACILLUS PENTOSUS* KCA1 FROM NIGERIA

After the post-doctoral/research fellowship in Canada, I returned and accepted a faculty position at the Department of Medical Laboratory Science, University of Benin. In 2010, I applied for and obtained a research grant from the Third World Academy of Science (TWAS) to determine the full genome sequence of *Lactobacillus plantarum* KCA1 and to set up Genomics Research Laboratory Unit (GRU) at the University of Benin. The genome description of *L. plantarum* KCA1 has been published (Anukam et al. 2013). The complete genome map is shown in Figure 7.5.

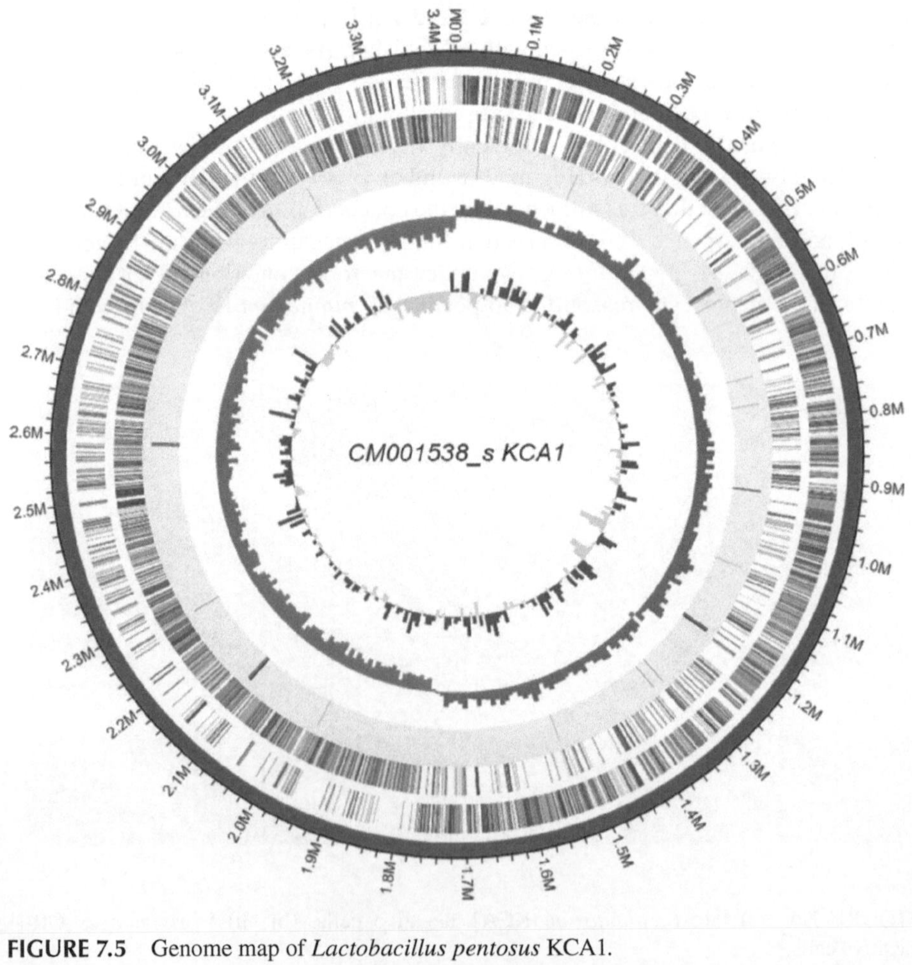

FIGURE 7.5 Genome map of *Lactobacillus pentosus* KCA1.

The reason for absence of gfp positive cells in *Lactobacillus pentosus* KCA1 strain was revealed after sequencing as the strain was found to be devoid of plasmids. It was at this stage that I had a sigh of relief. On further investigation, the *Lactobacillus pentosus* KCA1 genome contains a protein with 324 base pairs known as plasmid maintenance system killer protein (K07334, #PfamID (PF05015.6); Annotation-Plasmid_killer; e-value 1.3e-09; COG class-COG3549. The amino acid sequence of the KCA1 protein is given below:

MKYKAKSAKFQKQLTEPRLRMRTFGASVARRIGARIDEIAAAKTLTDVSRFPPSRLHMLSGDRTGLFAVDVSANYRLLFAGFDKQQQTLDPSEVVAVVFVDIEDYH

Similar organisms with plasmid killer protein are found in *Lactobacillus antri* DSM 16041 and *Lactobacillus brevis* ATCC 367. The plasmid killer protein in *Lactobacillus pentosus* KCA1 may have inactivated or killed the gfp plasmid pSMJJ8, and it could not be electroporated into the KCA1 genome.

One may ask why *Lactobacillus plantarum* KCA1 changed its name to *Lactobacillus pentosus* KCA1? Initially, based on the partial 16S rRNA gene sequences, the bacterium was identified as *L. plantarum*. However, we renamed the isolate as *Lactobacillus pentosus* KCA1 on the basis of the gene sequences of recA (recombinase A), dnaK (heat shock protein HSP70) and pheS (phenylalanyl-tRNA synthase alpha subunit) (Bringel et al. 2005), as these genes were able to discriminate all the species and subspecies of *L. plantarum* and *L. pentosus* (Huang et al. 2010; Naser et al. 2007).

7.10 CONCLUSION

The journey that led to the isolation, determination of probiotics properties and sequencing of the whole genome of *Lactobacillus pentosus* KCA1 has been inspiring and at the same time rewarding. The KCA1 strain is now used as a reference strain in some North American and European research laboratories as several research studies are currently going on in order to understand the mechanisms of the unique characteristics of *Lactobacillus pentosus* KCA1. In addition, *Lactobacillus pentosus* KCA1 strain is now used as a probiotic in the treatment of BV/vulvovaginal candidosis (Oerlemans et al. 2020), skin acne (Lebeer et al. 2019) and control of armpit body odor (Onwuliri et al. 2021), based on preliminary study showing effects of deodorants/antiperspirants on bacterial communities (Anukam et al., 2020) The *Lactobacillus pentosus* KCA1 has been licensed to Winclove Probiotics in the Netherlands. The potential applications of *Lactobacillus pentosus* KCA1 are yet to be fully harnessed as more in vitro and in vivo studies are currently going in some laboratories around the world.

REFERENCES

Antonio, M.A., Hawes, S.E., Hillier, S.L. (1999). The identification of vaginal *Lactobacillus* species and the demographic and microbiologic characteristics of women colonized by these species. *Journal of Infectious Diseases* 180:1950–1956.

Anukam, K.C. (2004). Probiotic functions of *Lactobacilli* – a review. *Journal of Medical Laboratory Science* 13(2):8–18.

Anukam, K.C., Koyama, T.E. (2007). Bile and acid tolerance of *Lactobacillus plantarum* KCA-1: a potential probiotical agent. *International Journal of Dairy Science* 2(3):275–280.

Anukam, K.C., Macklaim, J.M, Gloor, G.B, et al. (2013). Genome sequence of *Lactobacillus pentosus* KCA1: vaginal isolate from a healthy premenopausal woman. *PLoS One* 8(3):e59239. doi: 10.1371/journal.pone.0059239.

Anukam, K.C., Mmewurum, V., Agbakoba, N.R. (2020). Axillary microbiota compositions from men and women in a Tertiary Institution-South East Nigeria: effects of deodorants/antiperspirants on bacterial communities. BioRxivPreprint. https://doi.org/10.1101/2020.03.11.986364.

Anukam, K.C., Osazuwa, E.O., Ahonkhai, I., Ngwu, M., Osemene, G., Bruce, A.W., Reid, G. (2006b). Augmentation of antimicrobial metronidazole therapy of bacterial vaginosis with oral probiotic *Lactobacillus rhamnosus* GR-1 and *Lactobacillus reuteri* RC-14: randomized, double-blind, placebo-controlled trial. *Microbes and Infection* 8(6):1450–1454.

Anukam, K.C., Osazuwa, E.O., Ahonkhai, I., Reid, G. (2006c). *Lactobacillus* vaginal microbiota of women attending a reproductive health care service in Benin City, Nigeria. *Sexually Transmitted Diseases*, 33(1):59–62.

Anukam, K.C., Osazuwa, E.O., Mbata, T.I., Ahonkhai, I.N. (2004a). Increased incidence of pertussis and parapertussis in HIV-1-positive adolescents vaccinated previously with whole-cell pertussis vaccine. *World Journal of Microbiology and Biotechnology*, 20(3):231–234.

Anukam, K.C., Osazuwa, E.O., Osadolor, H.B., Bruce, A.W., Reid, G. (2008). Yogurt containing probiotic *Lactobacillus rhamnosus* GR-1 and L. reuteri RC-14 helps resolve moderate diarrhea and increases CD4 count in HIV/AIDS patients. *Journal of Clinical Gastroenterology* 42(3):239–243.

Anukam, K.C., Osazuwa, E., Osemene, G.I., Ehigiagbe, F., Bruce, A.W., Reid, G. (2006a). Clinical study comparing probiotic *Lactobacillus* GR-1 and RC-14 with metronidazole vaginal gel to treat symptomatic bacterial vaginosis. *Microbes and Infection* 8(12–13):2772–2776.

Anukam, K.C., Osazuwa, E.O., Reid, G. (2005). Improved appetite of pregnant rats and increased birth weight of newborns following feeding with probiotic *Lactobacillus rhamnosus* GR-1 and *L. fermentum* RC-14. *Journal of Applied Research* 5(1):46–52.

Anukam, K.C., Osazuwa, E.O., Reid, G. (2006d). Knowledge of probiotics by Nigerian clinicians. *International Journal of Probiotics and Prebiotics* 1(1):57–62.

Anukam, K.C., Osazuwa, E.O., Reid, G., Katsivo, M.N. (2004b). Receptivity for probiotic products among pre-menopausal female students in an African university. *Sexually Transmitted Diseases* 31(8):460–464.

Anukam, K.C., Osazuwa, E.O., Reid, G., Ozolua, R.I. (2004c). Feeding probiotic strains Lactobacillus rhamnosus GR-1 and *Lactobacillus fermentum* RC-14 does not significantly alter hematological parameters of Sprague-Dawley rats. *HAEMA Journal of Hellenic Society for Haematology* 7(4):497–501.

Anukam, K.C., Reid, G. (2007a). *Lactobacillus plantarum* and *Lactobacillus fermentum* with probiotic potentials isolated from the vagina of healthy Nigerian women. *Research Journal of Microbiology* 2(1):81–87.

Anukam, K.C., Reid, G. (2007b). Probiotics: 100 years (1907–2007) after Elie Metchnikoff's Observation. *Communicating Current Research and Educational Topics and Trends in Applied Microbiology*, A. Méndez-Vilas (Ed.), pp. 446–474.

Anukam, K.C., Reid, G. (2008). Effects of metronidazole on growth of *Gardnerella vaginalis* ATCC 1408, probiotic *Lactobacillus rhamnosus* GR-1 and a vaginal isolate *Lactobacillus plantarum* KCA. *Microbial Ecology in Health and Disease* 20(1):48–52.

Bradshaw, C.S., Morton, A.N., Hocking, J., Garland, S.M., Morris, M.B., Moss, L.M., Horvath, L.B., Kuzevska, I., Fairley, C.K. (2006). High recurrence rates of bacterial vaginosis over the course of 12 months after oral metronidazole therapy and factors associated with recurrence. *Journal of Infectious Diseases* 193(11):1478–1486.

Bringel, F., Castioni, A., Olukoya, D.K., et al. (2005) *Lactobacillus plantarum* subsp. *argentoratensis* subsp. nov., isolated from vegetable matrices. *International Journal of Systematic and Evolutionary Microbiology* 55:1629–1634.

Burton, J.P., Cadieux, P., Reid, G. (2003). Improved understanding of the bacterial vaginal microbiota of women before and after probiotic instillation. *Applied and Environmental Microbiology* 69:97–101.

Cohn, S.E., Clark, R.A. (2003). Sexually transmitted diseases, HIV, and AIDS in women. *Medical Clinics of North America* 87:971–995.

Falsen, E., Pascual, C., Sjoden, B., et al. (1999). Phenotypic and phylogenetic characterization of a novel Lactobacillus species from human sources: description of Lactobacillus inersspnov. *International Journal of Systematic Bacteriology* 49:217–221.

FAO/WHO (2001). Evaluation of health and nutritional properties of powder milk and live lactic acid bacteria. Food and Agriculture Organization of the United Nations and World Health Organization Report. Available at: http:// www.fao.org/es/ESN/probio/probio.htm.2001. Accessed April 20, 2020.

Huang, C.H., Lee, F.L., Liou, J.S. (2010) Rapid discrimination and classification of the *Lactobacillus plantarum* group based on a partial DNA K sequence and DNA fingerprinting techniques. *Antonie Leeuwenhoek* 97(3):289–296.

Lebeer, S., Oerlemans, E., Claes, I., et al. (2019). Topical cream with live lactobacilli modulates the skin microbiome and reduce acne symptoms. (A Pre-print). https://www.biorxiv.org/content/10.1101/463307v1. doi: 10.1101/463307.

Naser, S.M., Dawyndt, P., Hoste, B., et al. (2007) Identification of *Lactobacilli* by pheS and rpoA gene sequence analyses. *International Journal of Systematic and Evolutionary Microbiology* 57(Pt 12):2777–2789.

Oerlemans, E.F.M, Bellen, G., Claes, I., Henkens, T., Allonsius, C.N., Wittouck, S., Broek, M.F.L., Wuyts, S., Kiekens, F., Donders, G.G.G., Lebeer, S. (2020). Impact of a lactobacilli-containing gel on vulvovaginal candidosis and the vaginal microbiome. *Scientific Reports* 10:7976. doi: 10.1038/s41598-020-64705-x.

Onwuliri, V., Agbakoba, N.R., Anukam, K.C. (2021). Topical cream containing live lactobacilli decreases malodour-producing bacteria and downregulates genes encoding PLP-dependent enzymes on the axillary skin microbiome of healthy adult Nigerians. *Journal of Cosmetic Dermatology*. doi: 10.1111/jocd.13949. Advance online publication. https://doi.org/10.1111/jocd.13949.

Schwebke, J.R. (2001). Role of vaginal flora as a barrier to HIV acquisition. *Current Infectious Disease Reports* 3:152–155.

Sha, B.E., Zariffard, M.R., Wang, Q.J., et al. (2005). Female genital-tract HIV load correlates inversely with Lactobacillus species but positively with bacterial vaginosis and Mycoplasma hominis. *Journal of Infectious Diseases* 191:25–32.

Vasquez, A., Jakobsson, T., Ahrne, S., et al. (2002). Vaginal *Lactobacillus* flora of healthy Swedish women. *Journal of Clinical Microbiology* 40:2746–2749.

Wolf, B.W., Wheeler, K.B., Ataya, D.G., et al. (1998). Safety and tolerance of *Lactobacillus reuteri* supplementation to a population infected with human immunodeficiency virus. *Food and Chemical Toxicology* 36:1085–1094.

8 A Brief Review of Earth Microbiomes and Applications

Toochukwu E. Ogbulie
Federal University of Technology

Nwadiuto (Diuto) Esiobu
Applied Biotech International Nigeria Ltd
Florida Atlantic University

Ifeoma Enweani-Nwokelo
Nnamdi Azikiwe University

CONTENTS

8.1 Historical Perspectives 137
8.2 Definitions and Current Understanding of Microbiomes on Earth/Environment 139
8.3 Earth Microbiome Study 139
8.3.1 Protocol for Earth Microbiome Study 140
8.3.1.1 Sample Collection 141
8.3.1.2 Laboratory Procedures 141
8.3.1.3 Sequencing 142
8.3.1.4 Analysis of Biological Data 143
8.4 The Earth Microbiome and Its Applications 144
8.4.1 Application in Oil Exploration/Prospecting and Recovery 144
8.4.2 Application in Remediation and Sustainability (Terrestrial and Aquatic Environment) 145
8.4.3 Application in Fungal Diversity 152
8.4.4 Application in Agriculture and Food Production 152
References 156

8.1 HISTORICAL PERSPECTIVES

The earth is indeed a unique planet that provides utmost conditions that enable the existence of life. In comparison to our solar systems and other planetary systems, it is a special place to live. Indeed, the National Aeronautics and Space Administration (NASA) defined three planetary conditions that made the earth a habitable place. They include

DOI: 10.1201/9781003180241-8

the evolution of microbiome, energy sources for metabolic processes and the presence of complex organic compounds. Life on the planet earth has, however, been proved to have started about 3.7 billion years ago with only bacteria inhabiting in stromatolite as evidently proved by the presence of microbial fossil (Blaser et al., 2016). These cells vary in their reproductive cycle which most times are relatively short usually between 20 minutes of doubling time to 24 hours. Furthermore, due to DNA replication and with varying reproduction times, microbial evolution as well as easy adaptation to certain environmental factors (nutrients, water activity, temperatures, acidity, or humidity) can be enhanced. These microorganisms are the microbiota consisting of the total/community genome; the microbiomes are truly the biogeochemical engines on which the ecosystem functions rely upon for life on earth. Reports have evidently shown that the biosphere that exists today was facilitated by the oxidation event that occurred millions of years ago (Holland, 2006). This actually led to species diversification preceded by the existence of multi-cellular organisms and other domains of life (Figure 8.1a).

Historically, Figure 8.1a shows the microbial landmark in the evolution of the biosphere indicating the progressive nature of life from unicellular to multi-cellular microorganisms as well as formation of fungi, plants and animals kingdoms that have provided diverse habitats for microbial growth and proliferation. These plants and animals irrespective of the habitat where they exist can, just like the soil, host zillions of microbial species. In animals, for instance, the presence of these microbes is evident in their digestive systems where they form a dynamic ecosystem each of

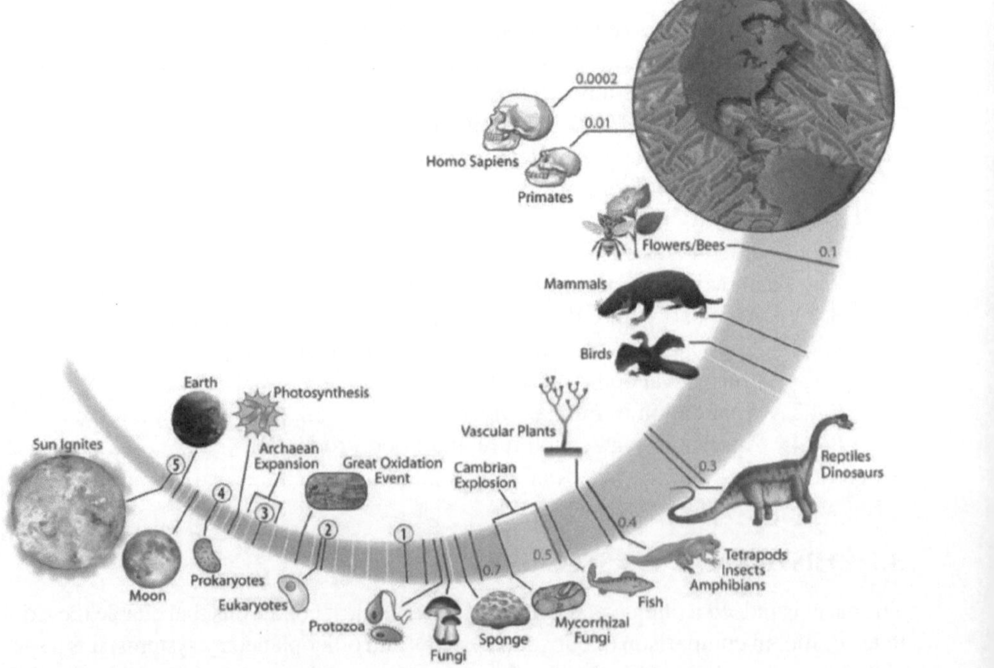

FIGURE 8.1 The biosphere and microbial evolution. (Source: Blaser et al., 2016.)

which possesses genes that constitutes the microbiome. A typical example is seen in the human gut microbiome which has been reported to house genes that are 100 times greater in number than that of human cells (Riedel et al., 2014; Ruairi et al., 2019). Nevertheless, reports made by Robinson et al. (2010) and Vogel et al. (2020) have shown that constant change in the resources of the host obviously played a role in the size and composition of these guest microbial communities. Consequently, it is obvious that the microbial community composition of the earth can change though depending on the variation in environmental factors (biotic and abiotic factors) at a point in time. Currently, a lot of studies (Caporaso et al., 2011, 2012; Gilbert et al., 2014; Thompson et al., 2017) on microbes and their environments are being documented, and the biological data generated have exceeded over years. This was made possible due to the advent of metagenomic analyses suggesting the need for great increase in capacity and analytical framework with relative ease.

8.2 DEFINITIONS AND CURRENT UNDERSTANDING OF MICROBIOMES ON EARTH/ENVIRONMENT

In 2001, Joshua Lederberg coined the term "microbiome" during which he conceptualized that the commensals, pathogenic and symbiotic organisms share our body space. The same is applicable to the organisms (bacteria and archaea) residing in an environmental niche or colonizing the earth. Further, the human microbiome was technically referred to as the complete genome of ecological community of microbes in and on the body surface of human. The earth microbiome on the other hand is referred to as the complete genome of the community of all microorganisms in the planet earth. The microbiome and the host, however, emerged during evolution as a synergistic unit from epigenetics and genetic characteristics, sometimes collectively referred to as a holobiont.

8.3 EARTH MICROBIOME STUDY

The increase in awareness of the microbes and its diverse nature and importance in the world is in contrast with the knowledge of its essential structure that forms the groundwork of the system. Despite the advent of DNA sequencing techniques, it is still very difficult to compare studies and draw global proposition or reach a global deduction on microbes and their life on earth. This is probably due to lack of recognized acceptable protocols and common analytical frameworks employed in carrying out research. However, metagenomic analysis of samples is currently done using protocols and analytical methods that are coordinated in studying the community DNA of the microbial community. Exact sequences are now used to amplify the target molecule, rather than using clustered units in an arbitrary order (operational taxonomic units (OTU)) to form axis of the dendrogram of mathematically defined taxa. This definitely grants opportunity to study the diversity of the microbial life, and the outcome will be based on biological data comprising of the DNA sequence data in a GenBank or reference database as well as deposition of biodata from new studies. This will consequently enhance complete characterization of earth's microbial diversity. However, the scientific understanding of the diverse nature of microbes in relation to their environment can be deduced from analyzing as many samples as possible so that the knowledge of the global distribution of microbes across environments can be documented. Therefore the primary target of the earth microbiome as reported by Thompson et al. in 2017 requires that the microbial diversity and prediction of functional profile from environmental samples should be studied. It is however, importantfor such task to be carried out by scientists across the globe. Further in their study, they carried out metagenomic analysis of microbial community samples and explored the pattern of microbial diversity using exact sequences instead of OTU. Their study showed that exact sequences are stable identifiers which enhance reusability because they can be compared to any 16S rRNA database. Nevertheless, the use of OTU is still applied in current study (Nayfach et al., 2020) on large-scale genomic resolved metagenomics to recover assembled genomes from diverse ecosystems representing broad phylogenetic diversity of culture-independent bacteria and archaea; perform metabolic modeling; and predict the functional profile and host-microbial linkages.

Sequencing of the 16S rRNA gene had been reported as a promising tool to survey the diversity of bacteria and archaea since it is a common taxonomic marker for both taxa (Robinson et al., 2010; Caporaso et al., 2011, 2012; Thompson et al., 2017). Several modifications have been employed to improve efficiency in situations where new phyla existence is suspected within the clades. Hence using structured categorical variables, the microbial diversity pattern in relation to its environment can be undertaken using the number of observed sequence tags and nestedness (α- and β-diversity) as well as genome predictions. This shows that there is obvious correlation between microbial community richness and environmental factors.

Indeed, the global project sites shown in Figure 8.2 are a clear insight on the extent the global team on earth microbiome had gone in sample collection (Blaser et al., 2016). They collected the total genome of the microbes (microbiome) in each sample and analyzed them using standard protocols with open data-sharing measures to generate reference database of microbes colonizing the earth. The target is, however, to archive the information from the project in a database to facilitate analysis resulting in global atlas of predictable functional activity/gene expression.

8.3.1 Protocol for Earth Microbiome Study

The standard protocols for sampling were basically designed following sample collection, DNA extraction, polymerase chain reaction (PCR), 16S rRNA and 18S rRNA amplification, NextGen sequencing and shotgun sequencing approaches/metagenomics, as well as sequence analysis and genome/protein annotations (Figure 8.3).

In bacteria and archaea as stated earlier, the major and most common taxa identifier that enhances the creation of evolutionary relationships among the taxa is the genes encoding the 16S rRNA small subunit of the ribosome. After obtaining the nucleotide sequences, researchers analyze the obtained sequences on comparison to nearly complete database of full length 16S rRNA gene sequences housed by the National Center for Biotechnology Information (NCBI). This can be accessed on their website at www.ncbi.nlm.nih.gov.

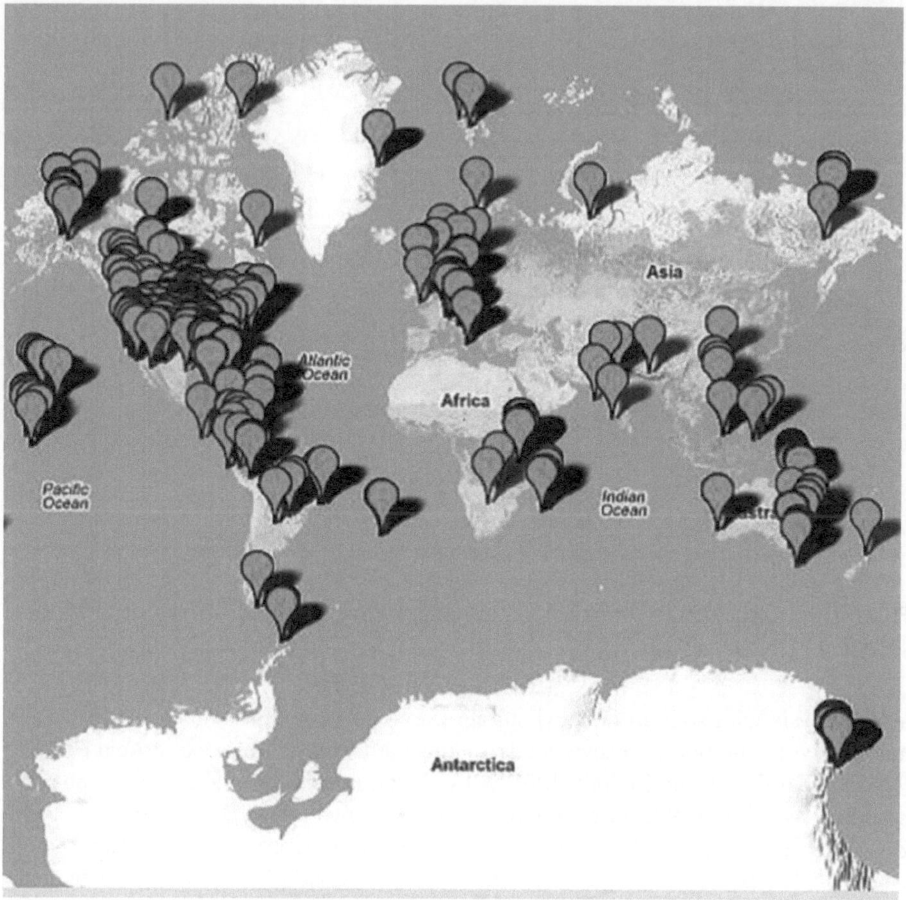

FIGURE 8.2 The Earth Microbiome Project sites with samples selected based on their position in environmental parameter gradients (Blaser et al., 2016).

8.3.1.1 Sample Collection

For efficient earth microbiome study to be accomplished, samples should be aseptically collected from various environments (aquatic and terrestrial) and from various regions (temperate; tropics) and vegetation using standardized collection protocols for comparable results to be obtained.

8.3.1.2 Laboratory Procedures

A series of procedures are required in carrying out the laboratory analyses. Microbes from the environmental samples are cultured, selected/identified and pure culture obtained. Varying purification processes are employed, although it depends on the type of sample. Isolates can equally be identified directly from the sample in which case the chromosomal DNA will be extracted from soil and the microbes too. The DNA extraction can be carried out directly following standard/modified procedures such as

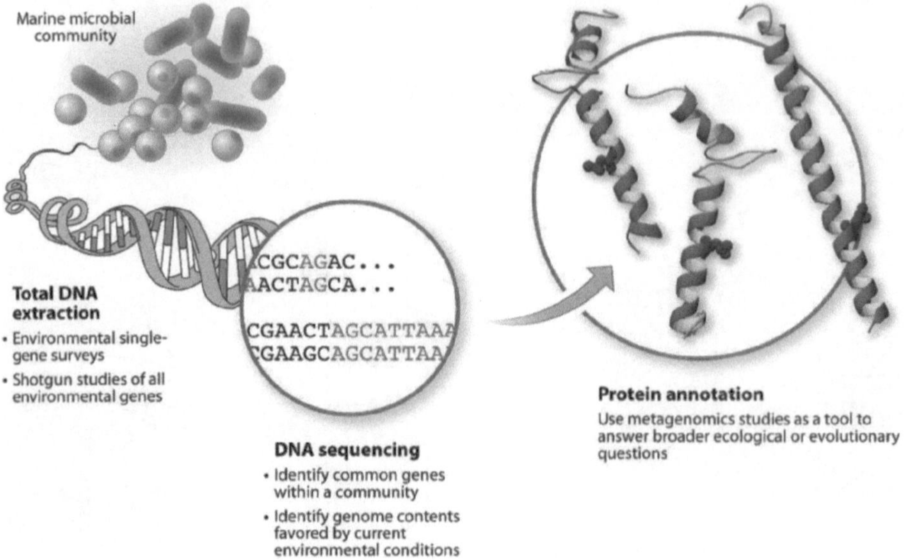

FIGURE 8.3 The standard protocols for sampling during earth microbiome study. (Source: Gilbert and Dupont, 2011.)

alkaline lysis with sodium dodecyl sulfate (SDS) and enzymes such as proteinase K and lysozyme and boiling lysis with Tris Ethylene Diamine Tetraacetic Acid (EDTA) (1X TE) (Sambrook and Russell, 2001) or by using filtration techniques involving commercially available DNA extraction kits made by many biotech companies such as the ZR Soil Microbe DNA Microprep (Zymo Research, CA, USA, www.zymoresearch.com), DNA extraction kit by Norgen Biotek Corp, Canada (www.norgenbiotek.com), Promega Corporation (www.worldwide.promega.com), Ultraclean soil DNA isolation kit – Cambio (www.cambio, co, uk), DNeasy PowerSoil Pro kit – QIAGEN (www.qiagen.com), etc. The type of kit to be used usually depends on the intended protocol to adopt as well as the type of sample/microorganism. This is usually followed by various amplification techniques as multiple displacement amplification, use of primers and PCR protocols following thermal cycler programs to increase DNA yield.

8.3.1.3 Sequencing

A metagenomic sample can be sequenced with different approaches depending on the biological query to be resolved. Sequences can be amplified randomly or specifically by targeting and amplifying highly conserved regions among consortium of microbes of interest which usually is the preferred approach. The 16S rRNA and 18S rRNA genes for bacteria and fungi are often respectively used as the taxonomical marker that targets specific genes for this purpose. This facilitates deep sequencing, enhancing the identification of culture-independent/rare species in a sample. However, deep sequencing approach does not enable whole genome assembly nor provide knowledge of microbial interaction with each other. Shotgun approach is rather a better alternative since it involves excision of the extracted DNA and sequencing of the random fragments, enabling whole microbial genome assembly and permitting induction of metabolic relationship.

Indeed, DNA sequence provides measurement of potential but not functional or activity profile of the microbiomes; other macromolecules (RNA, proteins, metabolites) are used instead to predict appropriate in situ activities of microbes with improved accuracy and integration together with spatial and temporal determination of gene expression (meta-transcriptomics) or synthesis of proteins (meta-proteomics) in microbes. This determines the major activities of the species individual functions in the ecological communities of the microbes. The complexity of the chemistry of microbiome is also made known by metabolomic approaches.

8.3.1.4 Analysis of Biological Data

Sequences are stored in database which serve as structured collection of information (biological) consisting of basic units called "records" or "entries". These entries/records consist of fields holding predefined data related to them in the form of a single file containing set of information. For instance, a record associated with a nucleotide sequence database typically contains information such as:

1. A contact name;
2. The input sequence with a description of the type of molecule;
3. The scientific name of the source organism from which it was isolated; and, often,
4. Literature citations associated with the sequence.

Nucleotide sequences in such case would represent the record/entry, while information such as the base count and origin, sequence length, etc. constitutes the field.

Databases therefore are designed to collect, archive, visualize and organize data to enable intelligent data analysis/description/interpretation, discovery, retrieval and invocation. They are also designed to ensure that different research institutes, research databases, data mining tools, software and ordinary end-user interact and function reciprocally with minimal restriction.

Furthermore, to analyze biological data, information will be retrieved from database since the records are organized in predetermined fields, which enable end-users to query on fields. This function is, however, made easier by use of algorithms, a software program designed to improve sequence-based database query (search) by increasing the speed, precision, accuracy and efficiency of identifying and making sense of similarities and/or dissimilarities from sequence alignments. Hidden Markov models, Smith-Waterman algorithm, Needleman-Wunch algorithm, etc. are typical examples of algorithms employed in sequence analysis. All these software are developed in such a way that it permits the integration of diverse data and tools under a common Graphic User Interface (GUI), sharing information and creation of powerful solutions useful in data archiving. Hence, biological data are analyzed by aligning sequences to retrieve information as follows:

1. Functional activity of a new gene/protein.
2. Structure of a new protein.

3. Location or preferred position of a protein.
4. Stability of a gene or protein.
5. Origin of a gene or protein.
6. Origin or phylogeny of an organelle.
7. Origin or phylogeny of an organism.

For earth microbiome project, identity of species can be retrieved at 97% lower limit from clustered OTUs using the UCLUST algorithm. Novel sequences discovered in certain situation and which cannot be mapped to existing taxonomy can deduce its taxa from the new sequences and collections of closely related known sequences by creating a phylogenetic tree. Alpha and beta diversity analysis in this case can be done using varying phylogenetic software packages such as Phyloseq package, PHYLIP, phyloT, etc. To obtain comparative gene families, functional prediction profiling of microbiome from the earth using 16S rRNA datasets is carried out using the PICRUSt and KEGG (Kyoto Encyclopaedia of Genes and Genomes) software.

Furthermore, when sequence reads are too short, useful information is usually not retrieved. This though depends on the sequence technology employed or the biological query. To solve this problem such that useful information on the species will be retrieved, the short reads will be assembled to construct whole genome. To also understand the metabolic relationships within a microbial metagenome, certain gene prediction tools (FragGeneScan, GENOMESCANGeMoMa, GENIUS, GENEID, etc.) will have to be employed to translate DNA sequences into amino acid sequences to observe the expression.

8.4 THE EARTH MICROBIOME AND ITS APPLICATIONS

For billions of years now, microorganisms have shaped our planet, although man has a profound influence on the biosphere which has been made known globally. Such influence include change in the use of land, change due to climatic conditions and industrialization, and response to rise in human population consequent upon profuse drift of man to urban settlements. The vast control of microbe across the biosphere thus requires cooperation among cross-disciplines to enhance understanding, prediction as well as harnessing of microbiome functions across the globe to help in eco-restoration and ecosystems modeling. This will enhance the development of new analytical and microbe-based approaches for improved prediction of response of the ecological community and its environment to significant challenges of our time.

8.4.1 Application in Oil Exploration/Prospecting and Recovery

In all shallow soil and sediment layers of the Earth, a variety of specialized microorganisms (hydrocarbon-oxidizing bacteria) exist. These microorganisms degrade short chain hydrocarbons easily especially when the concentration is low, maintaining active hydrocarbon-degrading population. Such continuous hydrocarbon supply areas are found around locations experiencing micro-seepages. In oil exploration, the relevant group of bacteria and fungi are basically hydrocarbon/methane oxidizers. The presence of these microorganisms enhances microbial prospecting for oil and gas (MPOG) which can be predicted with the help of earth microbiome study. In other words, exploring for the presence of oil and gas entails isolation of methane/hydrocarbon oxidizers on soil surfaces and analyzing/characterizing the isolates respectively using biochemical and molecular methods. Thereafter, increase in the number of viable cells representing the hydrocarbon/methane oxidizers becomes the indicator for the presence of oil/gas in the soil test samples. With earth microbiome study, suitable hitchhiker (microbes) for prospecting can therefore be selected and information documented in database on oil reservoir microbial communities.

Production of liquid hydrocarbon by method other than the conventional use of reservoir energy and reservoir re-pressurization with gas or water is termed oil recovery. It equally involves using microorganisms to extract the oil remains from oil reservoirs. Microbes in microbial-enhanced oil recovery are hydrocarbon-utilizing microorganisms and their ability to degrade hydrocarbon is the main criteria for their selection. These microorganisms are currently identified using bioinformatics tools where the functional activities of individual isolates are determined with available software [such as KEGG, see 8.3] for that purpose. Microbial-enhanced oil recovery processes which facilitate oil production involve multiple biochemical processes in crude oil. These processes involve the use of thermo-tolerant *aerobic* and facilitative anaerobic species of *Pseudomonas*, *Bacillus*, *Acinetobacter*, *Nocardia*, and *Rhodococcus*. These organisms, on the other hand, are earth microbiota, the microbiome of which can be studied to predict their functional ability. From the functional ability/metabolic profile (to be discussed in 8.4.2), the metabolic capabilities/gene products and functions are determined.

Microbial surfactants-producing gene from some earth microbiota can improve oil production/clean-up operations by contributing to removal of suspended debris and paraffin from near-well bore region. These microbial surface-active agents and bio-solvents also enhance oil production by reducing viscosity of heavy oil, thereby promoting emulsification of crude oil and altering relative permeability of reservoir rock to increase displacement of oil from the rock. Ogbulie and some group of researchers in 2017 (Okore et al., 2017a, b) isolated and identified microbes from soil with the gene for producing biosurfactants. Molecular identification was done by 16S rRNA gene sequencing (see 8.3) after screening for biosurfactant production using *β*-hemolysis test, oil displacement and emulsification index test, and several genera of bacteria were obtained. They include *Pseudomonas plecoglossicida*, *Pseudomonas aeruginosa* strain R873, *Enterobacter* sp. PD-P37, *Arcobacter trophiarum*, *Odoribacter splanchnicus* DSM 20712 and uncultured *Paenibacillus*, among others.

8.4.2 Application in Remediation and Sustainability (Terrestrial and Aquatic Environment)

Looking at the above outputs, the application of earth microbiome can be obviously felt in so many areas as bioremediation in both terrestrial and aquatic environment for enhanced sustainability. In biodegradation of petroleum products on polluted soil, bacteria are the most active agents that serve as primary degraders. Reports have shown, however, that they have in their genes the necessary enzymes that enhance hydrocarbon degradation. In other words, there is need to understand the relationships among pollutant degrading microbes in soil polluted crude oil. Therefore, for effective bio-engineered remediation to be achieved, the interaction effects of microbes on pollutants, as well as the reaction of inhabiting microbes (autochthonous) to certain geographical properties and chemical stress within the polluted site, must be put into consideration. Hence knowledge of the microbes present in the polluted sites will help researchers to understand the complex nature of microbial assemblage and establish their resilience and responses to pollutants. This is because previous reports have indicated that the best organisms that easily adapt to polluted or stressful environmental condition are mostly the indigenous microorganisms (Watanebe, 2001; Ogbulie et al., 2010, 2015; Blaise-Chikere et al., 2019). The reports of these authors have shown that metagenomic study of the polluted soils provided a better understanding of the syntrophic communities capable of eco-restoration. This will facilitate the establishment of the functional activities and interaction effects of the microbial communities. Currently, the earth microbiome study enables researchers that are into eco-restoration and biodiversity studies to use sequencing techniques such as next-generation sequencing (NGS) to identify microorganisms that have genes with conserved domains of unknown functions using new computational approaches as reported by Follows et al. in 2007 to predict function using high-throughput DNA synthesis and protein expression; to characterize microbes using scanning transmission X-ray microscopy (STXM), complementary metal-oxide semiconductor (CMOS), micro-computed tomography (microCT), nanoscale secondary ion mass spectrometry (NanoSIMS) and nanometer-scale stable isotope probing (NanoSIP); to establish robust hypothesis testing using transposon sequencing (TnSeq), numerical models; and to carry out the genome-centric quantitative analysis of microbiome using genome references catalogs, long read DNA sequencing, etc. as shown in Figure 8.4.

A typical example of remediation study where the interaction effect of microbes on soil was studied using genome-centric quantitative analysis (long read DNA sequencing and genome references catalogs were evidence in studies carried out by Ogbulie and Nwaokorie (2016)). The team carried out taxonomical analysis on short-term crude oil (Bonny Light crude oil) polluted agricultural soil with reads of sufficient

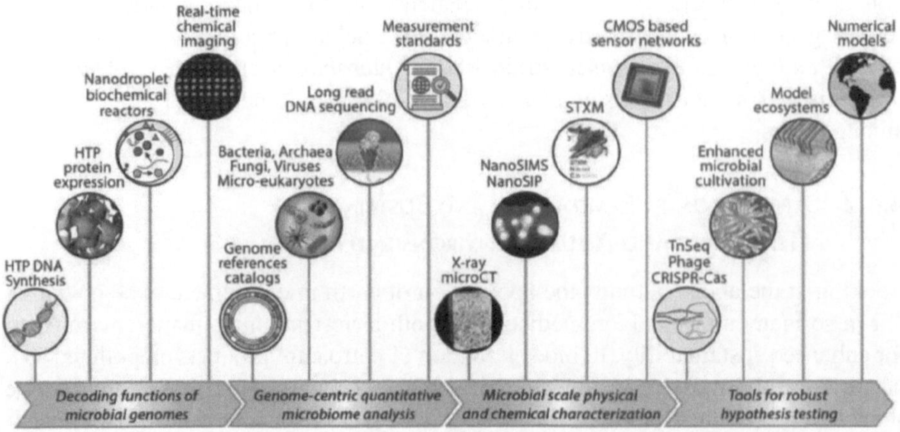

FIGURE 8.4 Decoding functions of microbial genomes. (Source: Follows et al., 2007.)

Q scores (>q30) revealing top kingdom, phyla, class, order and family classification of 100% bacteria kingdom (Figure 8.5-8). The charts represent the percentage taxonomical diversity of all the class of bacterial isolates based on Basic Alignment Search Tool (BLAST) output result of their study. Several researchers have reported bacterial isolates belonging to these phyla to be *Pseudomonas aeruginosa*, *P. fluorescence*, *Acidobacterium* sp., *Streptomyces* sp., *Nitrosovibrio* sp., *Burkholderia phymatum*, *Comamonas* sp., *Enterobacter* sp., *B. kururiensis*, *Klebsiella* sp., *Bradyrhizobium* sp., *Corynebacterium* sp., *Serratia* sp., *Acinetobacter baumani*, *Bacillus* sp., *Planococcus* sp., *Ralstonia* sp., *Norcadia* sp., *Lysobacter* sp., etc.

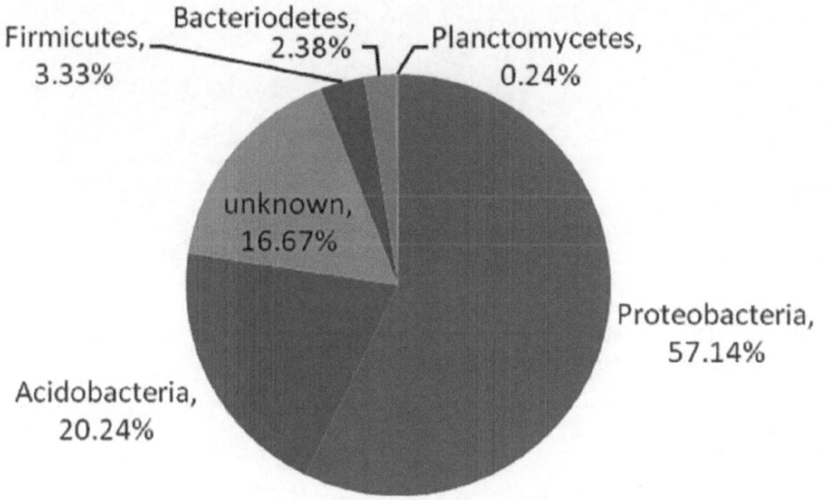

FIGURE 8.5 Bacterial phyla and its percentage diversity in crude oil polluted soil. (Source: Ogbulie and Nwaokorie, 2016.)

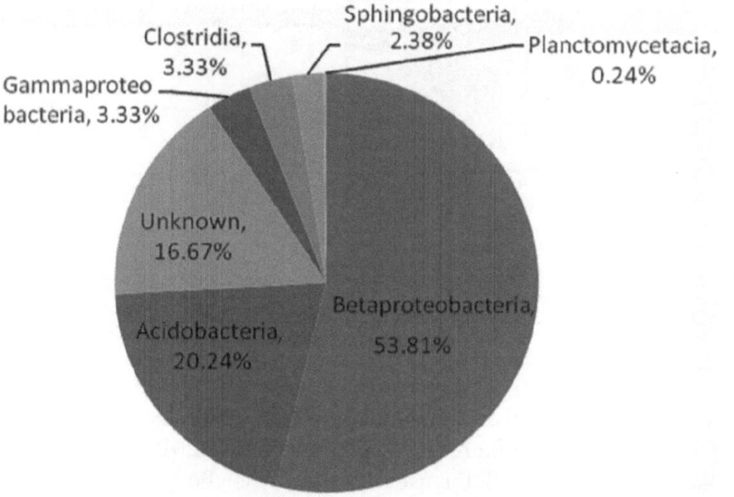

FIGURE 8.6 The diversity (%age) of class of bacterial species in crude polluted soil. (Source: Ogbulie and Nwaokorie, 2016.)

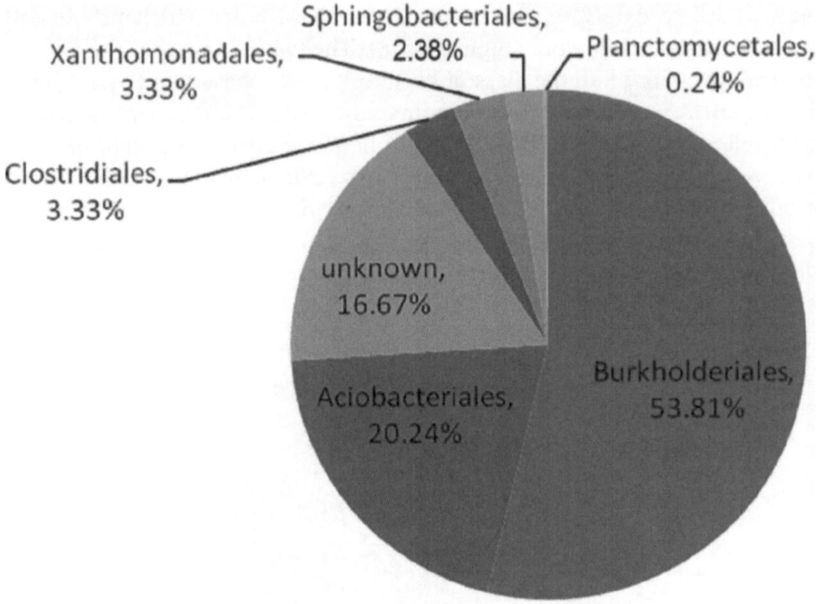

FIGURE 8.7 The top order classification of bacterial species and its diversity (%age) in polluted soil. (Source: Ogbulie and Nwaokorie, 2016.)

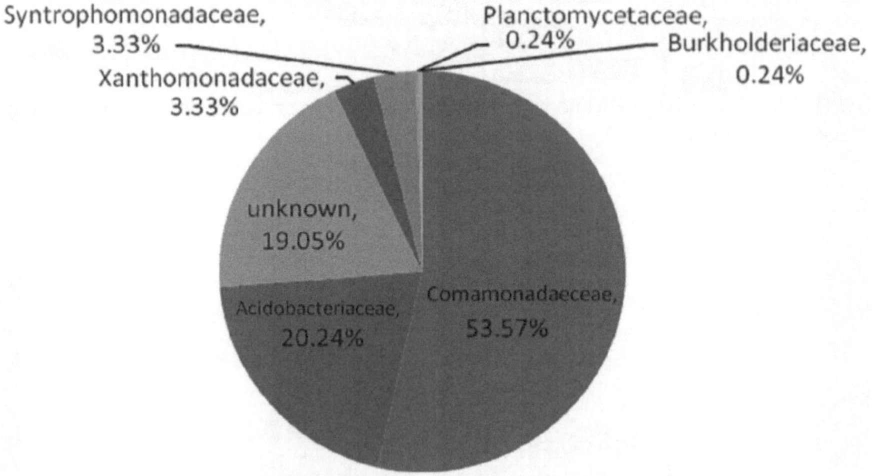

FIGURE 8.8 The taxonomical family of bacterial species and its diversity (%age) in polluted soil. (Source: Ogbulie and Nwaokorie, 2016.)

(Juhasz and Naidu, 2000; Wu et al., 2006; Kloos et al., 2006; Peng et al., 2008; Flocco et al., 2009; Obayori and Salam, 2010; Ogbulie et al., 2010, 2011, 2015; Ogbulie and Nwaokorie, 2016). In 2010, the use of some of the abovementioned microorganisms in alkane degradation was studied using monoculture and multiculture systems during soil clean-up operations (Ogbulie et al., 2010). Their findings recorded obvious petroleum hydrocarbon degradation evident through the result of alkane profile

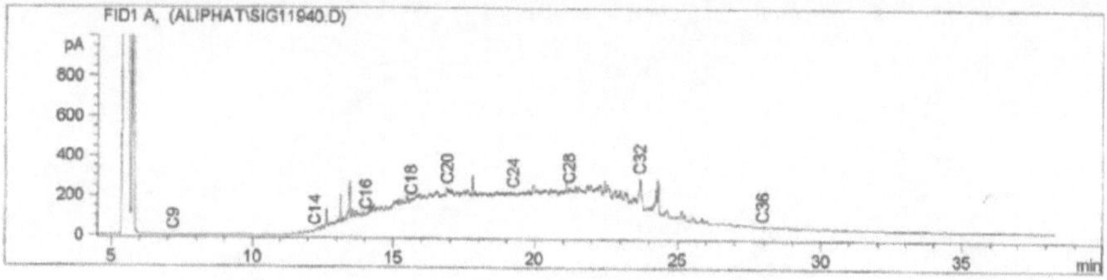

FIGURE 8.9 Gas chromatogram of alkanes degraded by only the microbial consortium present in the soil polluted with 100 mL of crude oil. (Source: Ogbulie et al., 2010.)

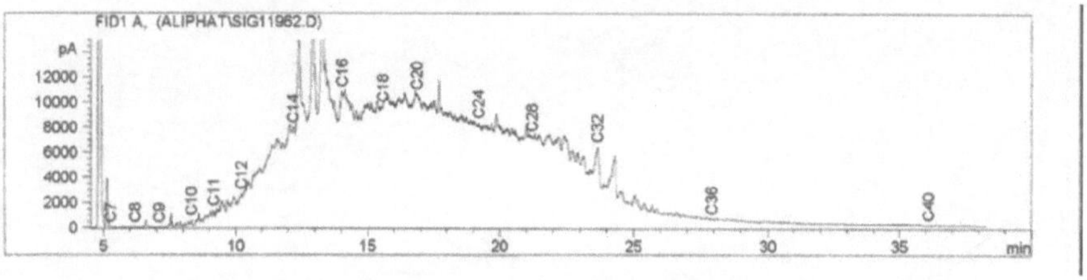

FIGURE 8.10 Gas chromatogram of all the alkane groups originally present in the test crude oil sample only (control) (Ogbulie et al., 2010, 2011).

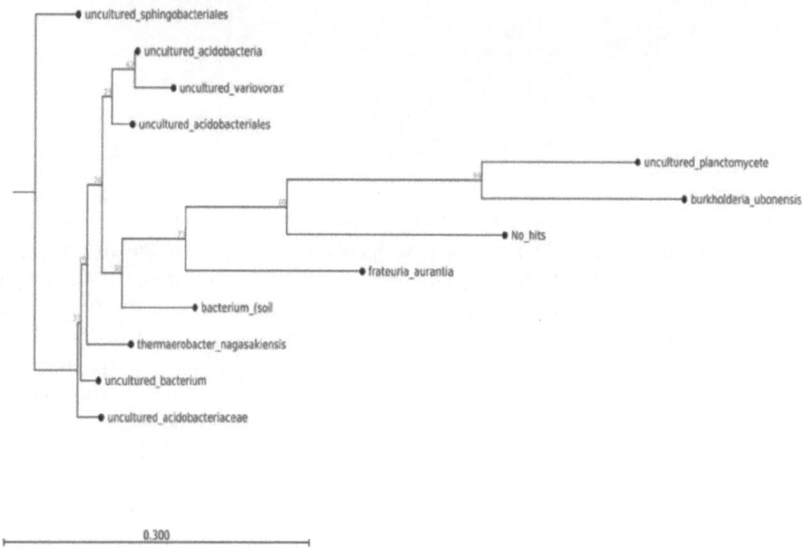

FIGURE 8.11 The dendrogram from cluster analysis depicting clades of taxonomical relationship. (Source: Ogbulie and Nwaokorie, 2016.)

(Figure 8.9 and 8.10) during microbiome study of crude oil polluted soil. The relatedness and the differences among the microbes resident in the test soil samples are, however, represented using the phylogenetic trees from that BLAST result hit based on similarity index and *e*-values (Figure 8.11).

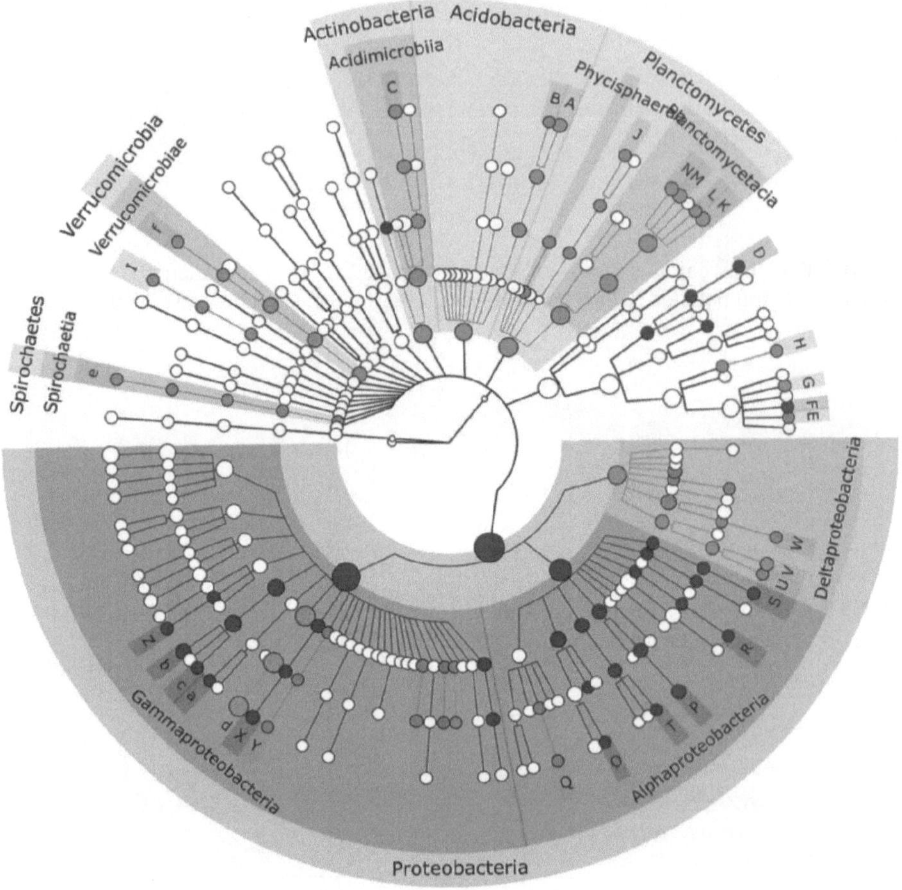

FIGURE 8.12 Taxonomic biomarkers representing the microbial phyla from pristine (grey) and soil polluted with crude oil (black). Legend: The alphabets A–Z represent the different phyletic subgroups where "A, subgroup 10; B, subgroup 23; C, Sva0996 marine group; D, Chryseolinea; E, Maritimimonas; F, Muricauda; G, Winogradskyella; H, Schleiferia; I, Truepera; J, Urania 1B 19 marine sediment group; K, Blastopirellula; L, Pir4 lineage; M, Rhodopirellula; N, Rubripirellula; O, Oceanicaulis; P, Parvibaculum; Q, Methyloceanibacter; R, Thalassospira; S, Sneathiella; T, Sphingobium; U, SEEP SRB1; V, Sva0081 sediment group; W, Haliangium; X, KCM B 112; Y, Arenicella; Z, C1 B045; a, Thioalkalispira; b, Fontimonas; c, Solimonas; d, Woeseia; e, Spirochaeta 2; f, Roseibacillus". (Source: Blaise-Chikere et al., 2019.)

Further, in the context of long-term oil pollution clean-up operation study, Blaise-Chikere et al. (2019) applied the use of earth microbiome studies to determine soil amenability to restoration by autochthonous bacteria and fungi present in crude oil polluted soil. The result of their study from both pristine and crude oil polluted soil is shown in Figure 8.12 striking difference between undisturbed and stressed soil microbiomes. The external and innermost rings of the figure thus represent the biomarker taxa and the taxonomic hierarchy colored according to the past assessment of notable crude oil pollution that occurred in the test soil. They also carried out DNA sequencing of the

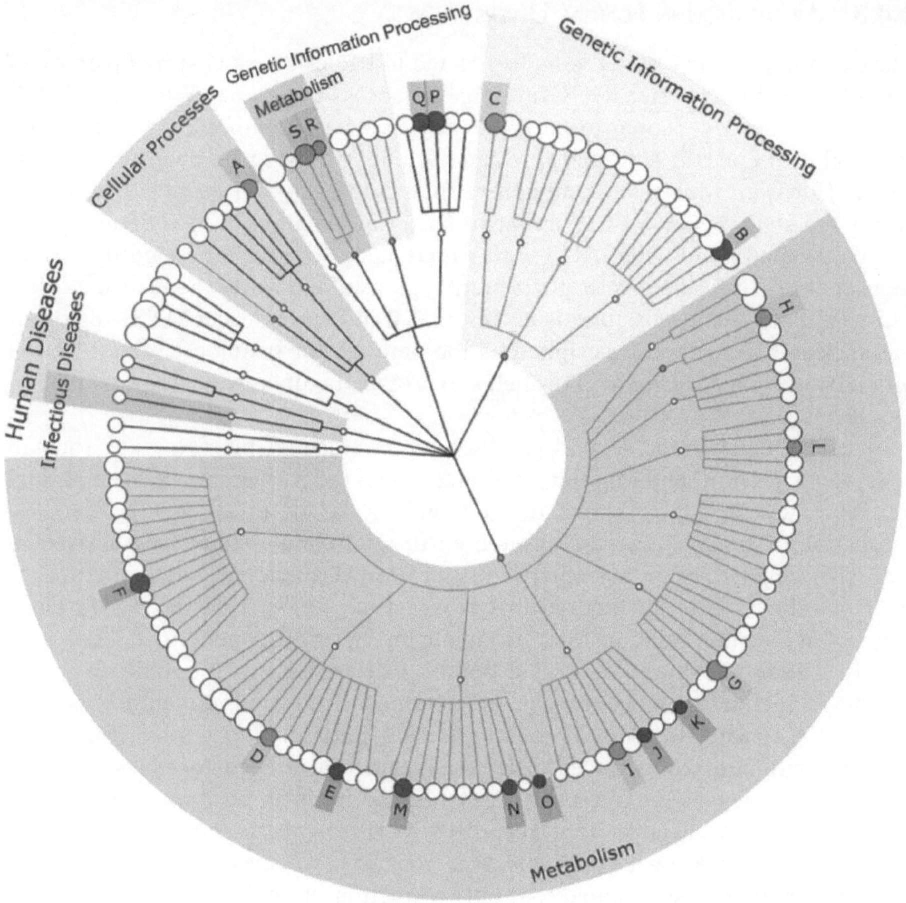

FIGURE 8.13 Functional prediction of using KEGG functional pathways with the circle size elucidating the biomarker abundance with colored pathway representing the study soil samples (undisturbed soil (pristine – grey) and soil polluted with crude oil (black)). The alphabets represent the different metabolic/functional profile. Legend: A, bacterial chemotaxis; B, chromosome; C, transcription machinery; D, lysine biosynthesis; E, tyrosine metabolism; F, glyoxylate and dicarboxylate metabolism; G, methane metabolism; H, protein kinases; I, pantothenate and CoA biosynthesis; J, riboflavin metabolism; K, vitamin B6 metabolism; L, prenyltransferases; M, benzoate degradation; N, naphthalene degradation; O, toluene degradation; P, membrane and intracellular structural molecules; Q, pores ion channels; R, amino acid metabolism; S, energy metabolism. (Source: Blaise-Chikere et al., 2019.)

soil community DNA, obtained the taxonomic profile data and subjected the sequence to PICRUSt analysis for functional predictions (proteins) (Figure 8.13) and further classified by KEGG ortholog clusters for genome comparison (https://www.genome.jp/kegg/kegg4.html). This played significant role in the identification of metabolic biomarker pathways associated with hydrocarbon degradation of polluted environment. Their findings showed therefore that the structure and function of surrounding ecosystem is a determining factor of the diversity of microbial community present therein.

8.4.3 Application in Fungal Diversity

The earth microbiome study is not restricted to bacterial species; biodiversity of fungi has also been extensively investigated across the globe (Wu et al., 2019). Fungal isolates are known to be frequently associated with ecosystems within the land of the earth, although few can be found growing on deep ocean sediments. However, the variety and general make up of mycobiome change widely across the study ecosystems. Similar to bacterial diversity, fungal diversity can be greatly facilitated using NGS with which many fungi can be identified from thousands of DNA samples from different regions of the world (Peay et al., 2016). This has been made possible using internal transcriber spacer (ITS) region of the nuclear rRNA operon designed as the barcode for taxonomic classification officially designed for fungi. This helps to provide taxonomic description of species levels.

Indeed, fungi are the second species-rich organisms which play key roles in ecosystems (Schmit and Mueller, 2007), although the pathogenic ones exists and has increased the global threat to food security (Hyde et al., 2018). However, estimation of fungal diversity through earth microbiome study have proved to play significant roles in systematic, resources, classification and prediction of their possible effects on food security (Wu et al., 2019). Further, the development of molecular relationships using phylogenetic/alignment trees will ultimately reveal uncultured fungal diversity. Through earth microbiome study, taxon-specific data sequences can be obtained from environmental samples, and this has paved way to new sources of data generation for global estimation of specie numbers. Nevertheless, the disadvantage in this context most times is that many sequences obtained cannot be aligned most times to already existing named fungi in GenBank. This according to Hawksworth and Lücking (2017) is attributed to unknown comprehensive geographic context of detected OTU. Their report therefore revealed that only a fraction of environmental sequences are deposited in GenBank by individual researchers, while the majority are placed on sequence read archives (SRA). However to solve this problem, global contribution by mycologists will be required to help understand fungal dimension and its implication for the welfare of man and its societies. Typical examples of fungal isolates obtained from environmental studies using culture-based method have been reported to include *Candida* spp., *Geotrichum candidum*, *Sepedonium* sp., *Aspergillus fumigatus*, *A. niger*, *A. nidulans*, *Cladophialophora carrionii*, *Gliocladium* sp., *A. terrens*, *Candida glabrata*, *Trichophyton tonsurans*, *Rhodotorula* spp., *Acremonium* sp., *A. versicolor*, *Cladosporium* sp., *Scopulariopsis brevicaulis*, etc. (Ogbulie et al., 2010, 2011, 2015; Ezekoye et al., 2018). They have been indicated to play obvious role in crude oil and heavy metal remediation of polluted soil.

8.4.4 Application in Agriculture and Food Production

In every grain of soil lie an excessive number of cells of microbes which are widely found in plants and animal tissues. Earth microbiome studies have played vital roles in understanding the regulation of the functions and activities of this living

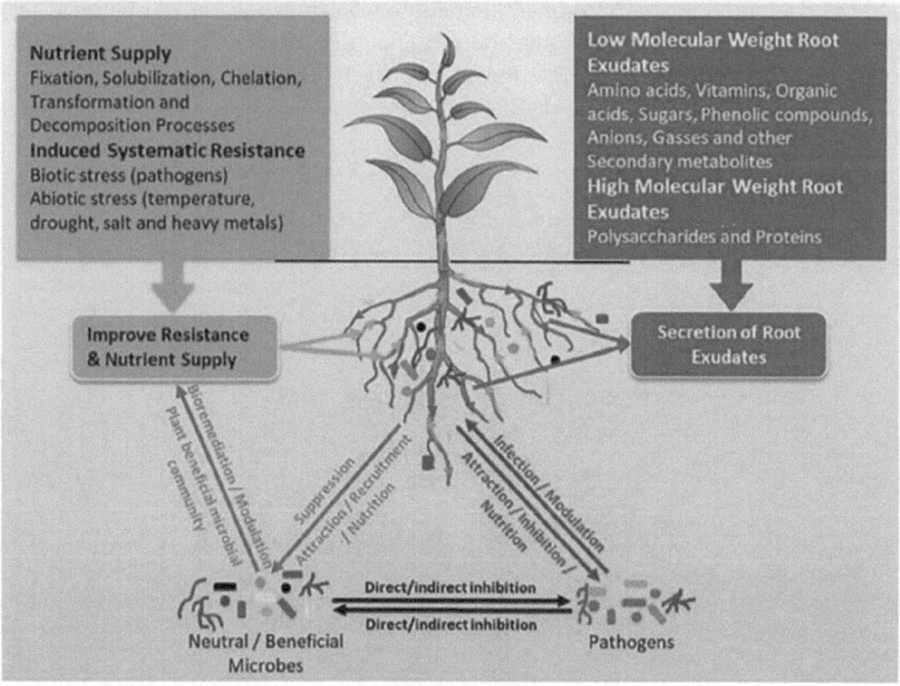

FIGURE 8.14 Plant functions and ecosystem well-being.

matter (Figure 8.14), as well as the physical and chemical phenomena involved (Andreote et al., 2014). Such functions could range from improved resistance and nutrient supply to secretion of low molecular weight exudates from plants while the microbes in turn enhance sustainable environmental management (Figure 8.15) through solubilization micro-macro elements, induced systemic resistance (ISR), antibiotics and enzyme secretion, quorum sensing, production of phyto-hormones, etc. (Mishra et al., 2016). Indeed, the components of these microbes in plant form communities with noticeable differences such as those inhabiting the rhizosphere (Figure 8.16), phylosphere and endosphere (Figure 8.17) of plants. However, studies on plant-microbe interaction have shown that most plant-related niches have significant role to play which in addition to already mentioned roles are beneficial to the health of plant and its development (Mendes et al., 2013; Philippot et al., 2013). These additional roles span through control of plant genetic structure, health and interaction with other plants, as well as how microbiome of the roots influences the cellular and microbial communities and ecosystem levels of plants (Figure 8.15).

Consequently, earth microbiome study can be harnessed to understand how microbes affect agricultural productivity through assisting and controlling of availability of nutrients as well as promoting tolerance to stressful conditions in the environment (Ferrara et al., 2012; Kawamura et al., 2013). This study can also help to understand how environmental conditions such as the use of chemical fertilizer in agricultural practices and the presence of pathogens can change plant-community composition and how the plants exert the selection of microbiomes (Wallis et al., 2010; Mendes et al., 2011; Rachid et al., 2013). It can also provide a

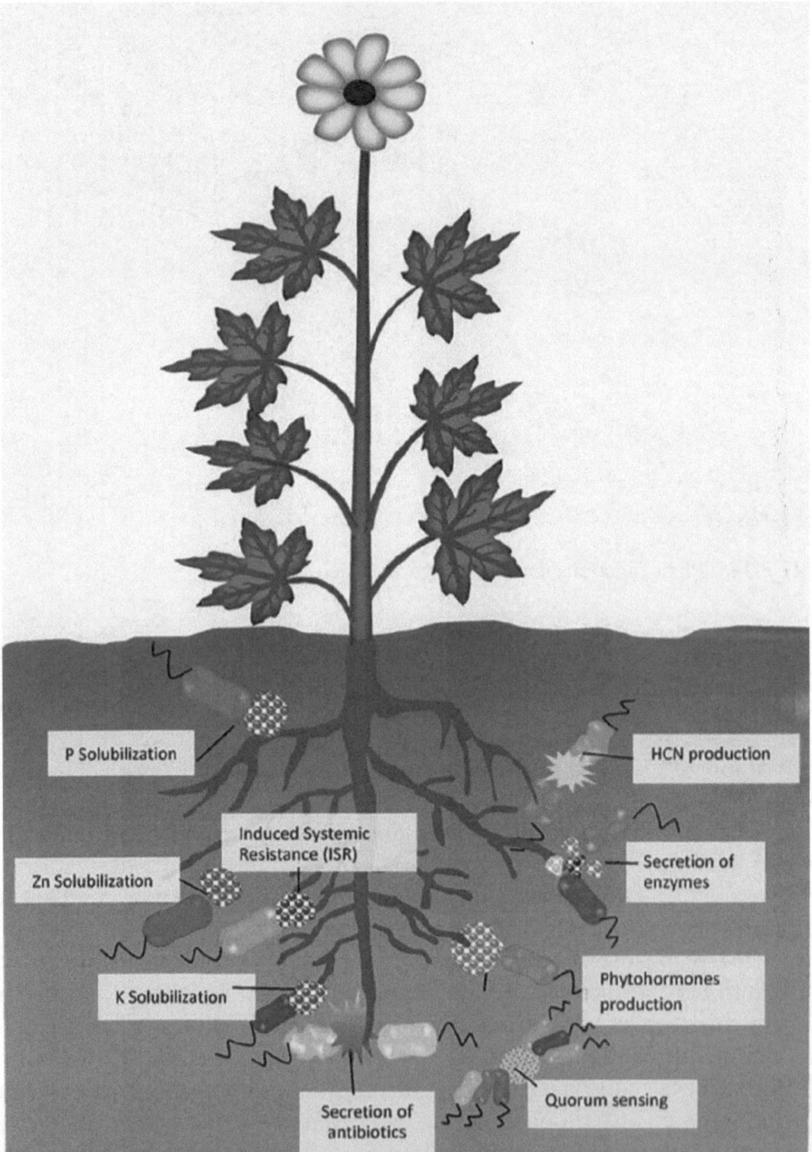

FIGURE 8.15 Beneficial soil microbes and their role in sustainable environmental management. (Source: Mishra et al., 2016.)

good understanding of bacterial and fungal interaction in soil which can encourage drastic reduction in soil fertilizer usage in agricultural practices and equally give a guide to the knowledge about future microbiome that will be able to suppress the development of pathogens in plants for enhanced productivity/high yield in food production thus providing solutions to global challenges of food security.

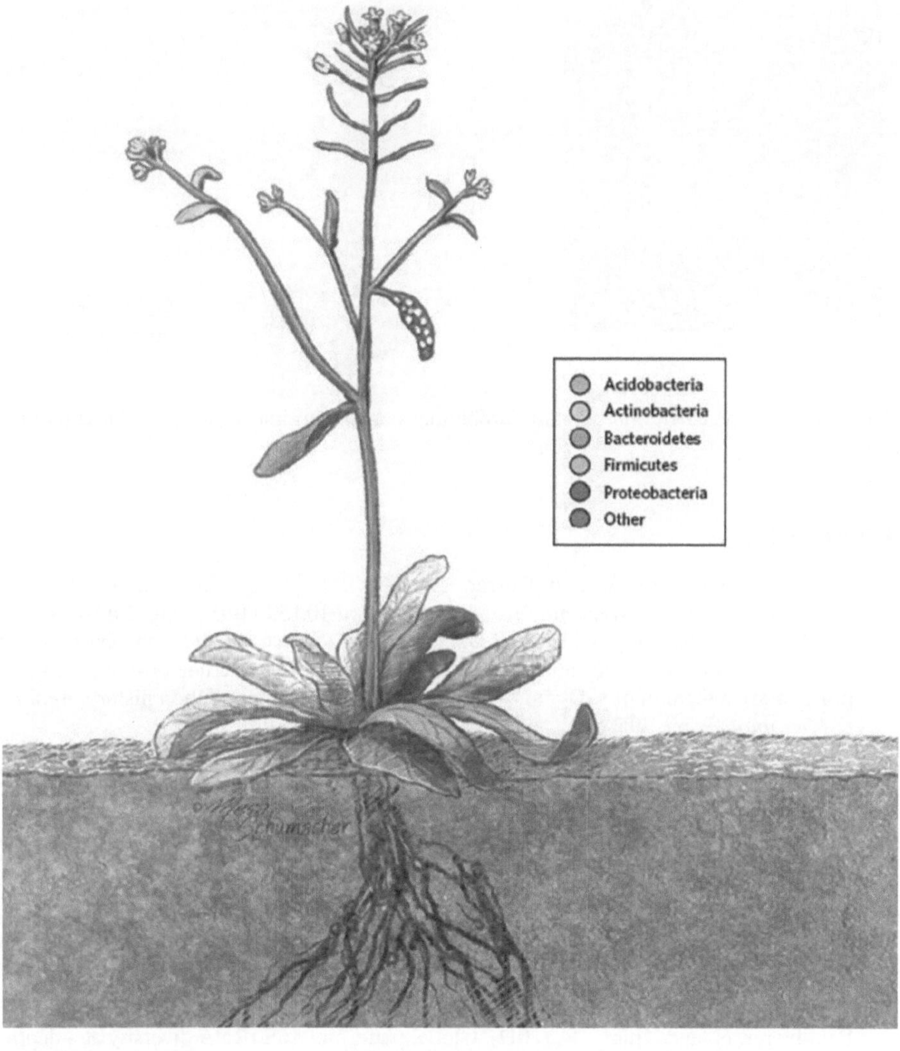

FIGURE 8.16 The plant root within the rhizosphere in interaction with microbiota, https://www.the-scientist.com/features/how-manipulating-the-plant-microbiome-could-improve-agriculture-30124.

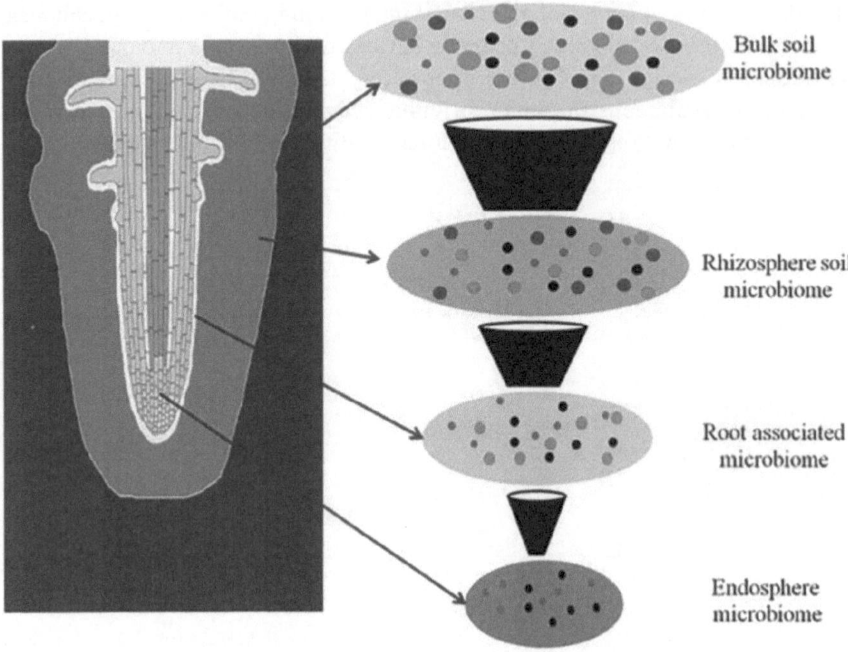

FIGURE 8.17 The earth microbiota/microbiomes across multiple scales in soil and root of plants.

REFERENCES

Andreote, F. D., Gumiere, T. and Durrer, A. (2014). Exploring interactions of plant Microbiomes. *Scientia Agricola* 71(6). https://doi.org/10.1590/0103-9016-2014-0195.

Blaise-Chikere, C., Mordi, I. J., Chikere, B.O., Selvarajan, R., Ashafa, T.O. and Obieze, C.C. (2019). Comparative metagenomics and functional profiling of crude oil-polluted soils in Bodo West Community, Ogoni, with other sites of varying pollution history. *Annals of Microbiology* 69: 495–513.

Blaser, M. J., Cardon, Z. G., Cho, M. K., Dangl, J. L., Donohue, T. J., Green, J. L., Knight, R., Maxon, M. E., Northen, T. R., Pollard, K. S. and Brodie, E. L. (2016). Toward a predictive understanding of Earth's microbiomes to address 21st century challenges. *ASM mBio* 7(3): e00714–16.

Caporaso, J. G., Lauberb, C. L., Walters, W. A., Berg-Lyons, D., Huntley, J., Fierer, N., Owens, S.M., Betley, J., Fraser, L., Bauer, M., Gormley, N., Gilbert, J. A., Smith, G. and Knight, R. (2012). Ultra-high-throughput microbial community analysis on the Illumina HiSeq and MiSeq platforms. *International Society for Microbial Ecology* 6: 1621–1624. www.nature.com/ismej.

Caporaso, J. G., Lauberb, C. L., Walters, W. A., Berg-Lyons, D., Lozuponea, C. A., Turnbaughd, P. J., Fierer, N. and Knight, R. (2011). Global pattern of 16S rRNA diversity at a depth of millions of sequences per sample. *Proceedings of the National Academy of Sciences* 108(1): 4516–4522. www.pnas.org/lookup/suppl/doi:10. 1073/pnas.1000080107/-/DCSupplemental.

Ezekoye, C. C., Chikere, C. B. and Okpokwasili, G. C. (2018). *In – situ* Bioremediation of Petroleum polluted site by enhanced natural attenuation. *Sustainable Chemistry and Pharmacy* 10: 148–152.

Ferrara, F. I. S., Oliveira, Z. M., Gonzales, H. H. S., Floh, E. I. S. and Barbosa, H. R. (2012). Endophytic and rhizospheric enterobacteria isolated from sugar cane have different potentials for producing plant growth-promoting substances. *Plant and Soil* 353: 409–417.

Flocco, C. G., Gomes, N. C. M., Cormack, W. M. and Smalla, K. (2009). Occurrence, diversity of naphthalene dioxygenase genes in soil microbial communities from the Maritime Antarctica. *Environmental Microbiology* 11: 700–714.

Follows, M. J., Dutkiewicz, S., Grant, S. and Chisholm, S. W. (2007). Emergent biogeography of microbial communities in a model ocean. *Science* 315: 1843–1846. http://dx.doi.org/10.1126/science.1138544.

Gilbert, J. A. and Dupont, C. L. (2011). Microbial metagenomics: Beyond the genome. *Annual Review of Marine Science* 3: 341–371. https://doi.org/10.1146/annurev-marine-120709-142811.

Gilbert, J. A., Jansson, J. K. and Knigh, R. (2014). The Earth Microbiome project: successes and aspirations. *BMC Biology* 12: 69–74. http://www.biomedcentral.com/1741-7007/12/69.

Hawksworth, D. L. and Lücking, R. (2017). Fungal diversity revisited: 2.2 to 3.8 million species. *Microbiology Spectrum* 5(4): 1–17. doi: 10.1128/microbiolspec.FUNK-0052-2016.

Holland, H. D. (2006). The oxygenation of the atmosphere and oceans. *Phil Trans Royal Society B Biological Sciences* 361: 903–915. http://dx.doi.org/10.1098/rstb.2006.1838.

Hyde, K. D., Al-Hatmi, A., Andersen, B., Boekhout, T., Buzina, W., Dawson Jr., T. L., Eastwood, D. C., Gareth Jones, E. B., de Hoog, S., Kang, Y., Longcore, J. E., McKenzie, E. H. C., Meis, J. F., Pinson-Gadais, L., Rathnayaka, A. R., Richard-Forget, F., Stadler, M., Theelen, B., Thongbai, B. and Tsui, C. K. M. (2018). The world's ten most feared fungi. *Fungal Diversity*, 93, 161–194.

Juhasz, A. L and Naidu, R. (2000). Bioremediation of high molecular weight polycyclic aromatic hydrocarbons: a review of the microbial degradation of benzo[a]pyrene. *International Biodeterioration and Biodegradation* 45: 57–88. www.elsevier.com/locate/ibiod.

Kloos, K., Munch, J. C. and Schloter, M. (2006). A new method for the detection of alkane monooxygenase homologous genes (alkB) in soils based on PCR-hybridization. *Journal of Microbiological Methods* 66: 486–496.

Mendes, R., Garbeva, P. and Raaijmakers, J. M. (2013). The rhizosphere microbiome: significance of plant beneficial, plant pathogenic, and human pathogenic microorganisms. *FEMS Microbiology Reviews* 37: 634–663.

Mendes, R., Kruijt, M., Bruijn, I., Dekkers, E., Voort, M. van der, Schneider, J. H. M., Piceno, Y. M., DeSantis, T. Z., Andersen, G. L., Bakker, P. A. H. M. and Raaijmakers, J. M. (2011). Deciphering the rhizosphere microbiome for disease-suppressive bacteria. *Science* 332: 1097–1100.

Mishra, J., Prakash, J. and Arora, N. (2016). Role of beneficial soil microbes in sustainable agriculture and environmental management. *Climate Change and Environmental Sustainability* 4(2): 137–143. doi: 10.5958/2320642X.2016.00015.6.

Nayfach, S., Roux, S., Seshadri, R., Udwary, D., Varghese, N., Schulz, F., … Eloe-Fadrosh, E. A. (2020). A genomic catalog of Earth's microbiomes. *Nature Biotechnology* 1–16. https://doi.org/10.1038/ s41587-020-0718-6.

Obayori, S. O. and Salam, L. B. (2010). Degradation of polycyclic aromatic hydrocarbons: role of plasmids. *Science Resource Ess* 5: 4093–4106.

Ogbulie, T. E., Duru, C., and Nwanebu, F. C. (2015). Interaction effects of plants and indigenous micro-organisms on degradation of N-alkanes in crude oil contaminated agricultural soil. *Journal of Ecosystem and Ecography* 5: 2–16. http://dx.doi.org/10.4172/2157-7625.1000166.

Ogbulie, T. E. and Nwaokorie, F. O. (2016). Molecular diversity of microbes with probable degradative genes in agricultural soil contaminated with bonny light crude oil. *Journal of Ecosystem and Ecography* 6: 1–9. http://dx.doi.org/10.4172/2157-7625.S5-002.

Ogbulie, T. E., Nwigwe, H. C., Iwuala, M. O. E. and Okpokwasili, G. C. (2010). Study on the use of Monoculture and Multispecie systems in Bioaugumentation of crude oil Contaminated Agricultural soil. *Nigerian Journal of Microbiology* 24(1): 2160–2167. www.nsmjournals.org.

Ogbulie, T. E., Nwigwe, H. C., Okpokwasili, G. C. and Iwuala, M. O. E. (2011). Comparative study on the effect of symbiotic interaction between plants and non-indigenous isolates on crude oil remediation. *Analele Universitatii din Oradea – Fascicula Biologie* 18(1): 15–22. http://www.bioresearch.ro/revistaen.html.

Okore, C. C., Nwaehiri, L. U., Mbanefo, O. N., Ogbulie, T. E., Ugenyi, A. U., Ogbuka, I. B., Ejele, A. E. and Okwujiako, I. A. (2017a). Study on microbial diversity of biosurfactant producing bacteria from contaminated environmental samples. *International Journal of Advanced Research* 5(6): 1387–1396.

Okore, C. C., Nwaehiri, L. U., Mbanefo, O. N., Ogbulie, T. E., Ugenyi, A. U., Ogbuka, I. B., Ejele, A. E. and Okwujiako, I. A. (2017b). Study on the production and Antibacterial activities of biosurfactant produced from some bacterial species. *International Journal of Advanced Research* 5(7): 581–586.

Peay, K. G., Kennedy, P. G and Talbot, J. M. (2016). Dimensions of biodiversity in the Earth mycobiome. *Nature Reviews* 14: 434–447.

Peng, R., Xiong, A., Sue, Y., Fu, X., Gao, F., Zhao, W., Tian, Y. and Yao, Q. (2008). Microbial biodegradation of polyaromatic hydrocarbons. *FEMS Microbiology Review* 32: 927–955.

Philippot, L., Raaijmakers, J. M., Lemanceau, P. and Putten, W. H. van der (2013). Going back to the roots: The microbial ecology of the rhizosphere. *Nature Reviews Microbiology* 11: 789–799.

Rachid, C. T. C. C., Santos, A. L., Piccolo, M. C., Balieiro, F. C., Coutinho, H. L. C., Peixoto, R. S., Tiedje, J. M. and Rosado, A. S. (2013). Effect of sugarcane burning or green harvest methods on the Brazilian Cerrado soil bacterial community structure. *PLoS One* 8: e59342.

Riedel, U., Schwiertz, A. and Egert, M. (2014). The stomach, small and large intestinal microbiome. *Advances in Molecular and Cellular Microbiology* 25: 1–19.

Robinson, C. J., Bohannan, B. J. M. and Young, V. B. (2010). From structure to function: the ecology of Host-Associated microbial communities. *Microbiology and Molecular Biology Reviews* 74(3): 453–476. doi: 10.1128/MMBR.00014-10.

Ruairi, C. R., Amee, R. M., Finlay, B. B. and Prendergast, A. J. (2019). The human microbiome and child growth first 1000 days and beyond. *Trends in Microbiology* 27(2), 131–147.

Sambrook, S. and Rusell, R. W (2001). *Molecular cloning: A laboratory manual* (3rd ed. Vol 1). New York: Cold spring harbor.

Schmit, J. P. and Mueller, G. M. (2007). An estimate of the lower limit of global fungal diversity. *Biodiversity Conservation* 16, 99–111.

Thompson, L. R., Sanders, J. G., McDonald, D., Fogliano, V., Jurburg, S. D. and Larsen, P. (2017). A communal catalog reveals Earth's multi-scale microbial diversity. *Nature* 551: 457–481. doi: 10.1038/nature24621.

Vogel, M. A., Mason, O. U. and Miller, T. E. (2020). Host and environmental determinants of microbial community structure in the marine phyllosphere. *PLoS One* 15(7): e0235441. https://doi.org/10.1371/journal.pone.0235441.

Wallis, P. D., Haynes, R. J., Hunter, C. H. & Morris, C. D. (2010). Effect of land use and management on soil bacterial biodiversity as measured by PCR-DGGE. *Applied Soil Ecology* 46: 147–150.

Watanbe, K., Kodoma, Y., Stutsubo, K. and Harayama, S. (2001). Molecular characterization of bacterial populations in petroleum-contaminated ground water discharge from undergoing crude oil storage cavities. Applied and Environmental Microbiology 66:4803-4809

Wu, B., Hussain, M., Zhanga, W., Stadler, M., Liua, X. and Xianga, M. (2019). Current insights into fungal species diversity and perspective on naming the environmental DNA sequences of fungi. *Mycology* 10(3): 127–140. https://doi.org/10.1080/21501203.2019.1614106.

Wu, C. H., Wood, T. K., Mulchandani, A. and Chen, W. (2006). Engineering plant microbe symbiosis for rhizoremediation of heavy metals. *Applied and Environmental Microbiology* 72(2): 1129–1134.

Ye, Y. and Doak, T. G. (2009) A parsimony approach to biological pathway reconstruction/inference for genomes and metagenomes. *PLoS Computational Biology* 5(8): e1000465. https://doi.org/10.1371/journal.pcbi.1000465.

Index

Note: **Bold** page numbers refer to tables and *italic* page numbers refer to figures.

absolute uterine factor infertility (AUFI) 88
acetate 38, 101, 105
acid tolerance 124
Acquired Immunodeficiency Syndrome (AIDS) 121–123
Akkermansia 38, 61
Akkermansia muciniphila 35, 37, 42, *61,* 62, 103
Alam, R. 106
Alang, N. 85
Allegretti, J.R. 85
alpha diversity/mean species distribution 80, 103
amyotrophic lateral sclerosis (ALS) 37
Anderson, J.L. 87
antibiotic-associated diarrhea (AAD) 59
Antonio, M.A. 123
Aroniadis, O.C. 90
artificial ecosystem selection of microbiome (AESM) 26
autoimmune disorders
  inflammatory bowel disease 40–41
  rheumatoid arthritis 40
  systemic lupus erythematosus 39–40
  type 1 diabetes mellitus 39

bacterial vaginosis (BV) 32, 42, 88, 89
  probiotics 62
  treatment and cure 127–128
bacteriocins 124
Bacteroidetes 34, 39
Bdellovibrio bacteriovorus 44
beneficial soil microbes, in sustainable environmental management *154*
Bercik, P. 111
beta diversity 80
Bifidobacterium longum 103
bioeconomy 11–12
biological data analysis, in Earth microbiome study 143–144
biosphere and microbial evolution 138, *138*
biosurfactant production 124, **129**
Blaise-Chikere, C. 149
blood-brain barrier (BBB) 46, 110
body mass index (BMI) 34
butyrate 101

Cammarota, G. 85
Citrobacter rodentium 90
Clostridium difficile infection (CDI) 58, 81, 87
computational biology 11
Corynebacterium spp. 102

deep sequencing approach 142
DeLong, K. 87
depression 35–36
diabetes mellitus 34
diarrheal disease in children 58–59
diet and gut microbiota 82–83
dietary fibers 82–83
diet-microbiome-epigenetics crosstalk 104–105, *106,* **107–108**
diet-microbiota-host interaction health consequences 82
DNA extraction 141
DNA methylation 105
DNA sequencing techniques 139
Dominguez-Bello, M.G. 103
donor screening, for FMT 83–85
dysbiotic oral microbiome 102

Earth Microbiome Project sites 140, *142*
Earth microbiomes
  across multiple scales in soil and root of plants *156*
  historical perspectives 137–139, *138*
  and human microbiomes 139
Earth microbiome study 139–140, *141*
  applications
    in agriculture and food production 153–156, *153–156*
    in fungal diversity 151–153, *152*
    in oil exploration/prospecting and recovery 144–145
    in remediation and sustainability 145–151, *146–151*
  protocols for 140, *142*
    biological data analysis 143–144
    laboratory procedures 141–142
    sample collection 141
    sequencing 142–143
endosphere microbiome 22
epigenetics 110–111
  mode of action associated with 100
Escherichia coli 90

fecal microbiome profiling 111
fecal microbiota delivery methods 85–86, *86*

fecal microbiota transplantation (FMT) 26, 41, 63, 111
administration/infusion route of 80–81
application of 86
in cardiometabolic disorders 87
for CDI management 87
complications 81
donor screening for 83–85
fecal microbiota delivery methods 85–86
future trends of 91
from healthy donor to unhealthy recipient 91, *92*
for IBD patients 87
limitations of 90–91
normobiosis restoration 86
purpose of 84
therapeutic uses of 81–82
fiber-rich diets 82, 83
Firmicutes 34
FMT *see* fecal microbiota transplantation (FMT)
Fu, B.C. 100

Gardnerella vaginalis 129, 130
Garza-González, E. 89
gastro-intestinal tract (GIT) 40
Gemella spp. 102
Gfp-*L. plantarum* KCA1 negative cells *131*
Gfp-*L. rhamnosus* GR-1 positive cells *131*
Gram-negative bacteria 36, 44
Gram-positive bacteria 72
growth kinetics, of *Lactobacillus plantarum* KCA1, in VDMP 129, *130*
gut-brain relationship 106
gut microbial homeostasis 80
gut microbiome 31–32, 46
gut microbiota-derived metabolites 101; *see also* short-chain fatty acids (SCFA)
gut microbiota nonbacterial members screening 84

Hawksworth, D.L. 152
high-fat diet (HFD) 83
histone deacetylase (HDAC) inhibition 105
Hullar, M.A. 100
human and Earth microbiome 139
Human Immunodeficiency Virus (HIV) 121–123
human intestinal microbial flora 99
human microbiome 5, 9, 11–12
from birth
anatomical sites of body *68*
early development in 75–77
intestinal microbiome 74–75
lung microbiome 75
microbial interaction pathway *70*
oral cavity of 73–74
skin composition 70–72
stomach *69,* 74–75
urogenital microbiome 72–73
microbial biotechnology 34
composition 28–29
correlations with human health 30
early development of 29–30
in medicine 42–43
human microbiome project 30, 80
Human Oral Microbiome Database (HOMD) 45
human stomach *69,* 74–75
Hvas, C.L. 86
hydrogen peroxide 124
hygiene hypothesis 57

indomethacin-induced gastric erosion *126*
inflammatory bowel disease (IBD) 40–41, 87, 90
restoration after feeding with GR-1/RC-14 *126*
in situ microbiome engineering 26
intestinal microbiome 74–75, 102
intestinal microbiota colonization 100
intestinal microbiota transplantation 89–90

Jones, B.P. 89
juvenile diabetes *see* type 1 diabetes mellitus (T1DM)

Kelly, C.R. 85
Kim, T.T. 87

laboratory procedures, in Earth microbiome study 141–142
lactic acid bacteria 123
lactic acids, production of 124
Lactobacillus 103
Lactobacillus pentosus 62
Lactobacillus pentosus KCA1
biosurfactant production 124
emergence and sequencing of *132,* 132–133
genome map of *132*
Lactobacillus plantarum 44
Lactobacillus plantarum KCA1
green fluorescence protein electroporation *131,* 131–132
growth kinetics in VDMP 129, *130*
Lactobacillus rhamnosus 58
Lactobacillus spp. 90
colonization in African/Nigerian women's vaginas 126–127
probiotic properties of 128–129, **129**
Lederberg, Joshua 139
Lee, C.H. 86
Lev-Sagie, A. 88
long-term oil pollution clean-up operation study 149
Lou Gehrig's disease *see* amyotrophic lateral sclerosis (ALS)
Lücking, R. 152

lung microbiome, human 75
Lupus *see* systemic lupus erythematosus (SLE)

Ma, D. 88
malnutrition 102, 103
Meighani, A. 85
metabolic disorders
amyotrophic lateral sclerosis 37
depression 35–36
neurologic disorders 35
obesity 34
Parkinson's disease 37–38
type 2 diabetes 34–35
metabolic syndrome 42
metabolomic approach 143
metagenome wide association study (MGWAS) 61
metagenomic analysis 139
metagenomics 2
academic-based institutions 8
animal, human and soil microbes 6
application of 3–4
challenges of 8–10
history and evolution of 4–6
human prospects of 11–12
products from 4
service 6–9
workflow process *4*
Metchnikoff, Elie 123
Methanobrevibacter smithiiobligate anaerobes 103
microbial biotechnology
artificial ecosystem selection of microbiome 26
autoimmune disorders
inflammatory bowel disease 40–41
rheumatoid arthritis 40
systemic lupus erythematosus 39–40
type 1 diabetes mellitus 39
biochemical strategy 27
human microbiome 34
composition 28–29
correlations with human health 30
early development of 29–30
in medicine 42–43
metabolic disorders
amyotrophic lateral sclerosis 37
depression 35–36
neurologic disorders 35
obesity 34
Parkinson's disease 37–38
type 2 diabetes 34–35
microbial-based strategy 26–27
microorganisms
gut microbiome 31–32
oral microbiome 30–31
skin microbiome 33
vaginal microbiome 32–33
molecular approach 27–28
plant microbiome 20, 28
collaborative functional capabilities 22
endosphere microbiome 22
functional interactions 22, **23–24**
microbial inoculants 25–26
phyllosphere microbiota 21
rhizosphere microbiota 21
in situ microbiome engineering 26
probiotics
metabolic syndrome 42
vaginal infection 41–42
skin microbiome applications
cosmetics and anti-aging 44–45
forensic applications 43
gut microbiome applications 46
oral microbiomes applications 45–46
therapeutic applications 43–44
microbial-derived metabolic products 100–101
microbial dietary metabolism 105
microbial-enhanced oil recovery processes 145
microbial fermentation, of dietary fibers 82
microbial genomes, decoding functions of *146*
microbial inoculants 25–26
microbial surfactants-producing gene 145
microbiome 2, 3, 99
academic-based institutions 8
challenges of 8–10
future perspectives 111–112
history and evolution of 4–6
human prospects of 11–12
immunity and epigenetic modification 105–106, 109–111
products from 4
reshaping and epigenetic modification **107–108,** *112*
microbiome-gut-brain (MGB) axis 36
microbiome-mediated dietary metabolic activities 100
microbiota
of Burkina Faso children 101
diet and gut 82–83
phyllosphere 21
primordial 103
rhizosphere 21; *see also* microbiota transplantation
microbiota-diet-epigenetics axis 111
microbiota-targeted drugs 111
microbiota transplantation
fecal (*see* fecal microbiota transplantation (FMT))
intestinal (*see* intestinal microbiota transplantation)
vaginal (*see* vaginal microbiota transplantation (VMT))

microorganism biotechnology
gut microbiome 31–32
oral microbiome 30–31
skin microbiome 33
vaginal microbiome 32–33
Mullish, B.H. 86
multi-omics approach 109

nasoduodenal route, of FMT 81
National Aeronautics and Space Administration (NASA) 137
neurologic disorders 35
next-generation genome sequencing procedures 110
next-generation sequencing (NGS) 2, 7
Nigeria
metagenomics (*see* metagenomics)
microbiomes (*see* microbiomes)
next-generation sequencing 2
probiotics (*see* probiotics)
Nigerian diet 59–60
non-gonococcal urethritis (NGU) 73
Nonoxynol-9 121
nucleotide sequence database information 143

obesity, metabolic disorders 34
Ogbulie, T.E. 145
omics-based technologies 109
OMICS technology 5
OpenBiome 84
operational taxonomic units (OTU) 139, 140
oral cavity microbiome 73–74
oral microbiome 30–31, 45–46
oral rehydration therapy (ORT) 58
Osazuwa, Emmanuel 122

Parkinson's disease (PD) 37–38
phyllosphere microbiota 21
phytobiome *see* plant microbiome
plant, functions and ecosystem well-being 153, *153*
plant growth-promoting rhizobacteria (PGPR) 25
plant internal tissue microbiome 22
plant microbiome 20, 28
collaborative functional capacity 22
endosphere microbiome 22
functional interactions 22, **23–24**
microbial inoculants 25–26
phyllosphere microbiota 21
rhizosphere microbiota 21
in situ microbiome engineering 26
polymerase chain reaction (PCR) 71
prenatal microbiome 101
Prevotella spp. 102
primordial microbiota 103
probiotic microorganism, characteristics 123–124
probiotic research, in Nigeria 121–123
preliminary studies 123–125, **125**
probiotics
bacterial vaginosis 62
current market in Nigeria 62–63
defined 123
diarrheal disease in children 58–59
human clinical studies 127–128
metabolic syndrome 42
properties of *Lactobacillus* species 128–129, **129**
safety
and issues **125**
studies 125–126
type 2 diabetes mellitus 60–62
vaginal infection 41–42
westernization of Nigerian diet 59–60

Qin, Y. 109
quantitative trait loci (QTL) 27–28

Reid, Gregor 123, 127
rheumatoid arthritis (RA) 40
Rhizosphere engineering 27
rhizosphere microbiota 21
Robertson, R.C. 102
Robinson, C.J. 139
root microbiome 26

sample collection, in Earth microbiome study 141
sequencing
in Earth microbiome study 142–143
of *Lactobacillus pentosus* KCA1 132–133
Sharma, M. 109
short-chain fatty acids (SCFA) 62, 100, 101, 105
shotgun approach 142–143
single-nucleotide polymorphisms (SNPs) 71
skin microbiome 33
cosmetics and anti-aging 44–45
forensic applications 43
gut microbiome applications 46
in human 70–72
oral microbiomes applications 45–46
therapeutic applications 43–44
Sonnenburg, E.D. 83
16S rRNA gene sequencing 140, 145
stool banking 84, 85, 91
Streptococcus thermophilus 103
super-organisms 68
synthetic genomes 11
synthetic microbial consortia (SMC) 47
systemic lupus erythematosus (SLE) 39–40
systems biology 11

Thompson, L.R. 140
transgenerational transmission, of genetic modifications *112*, 113

type 1 diabetes mellitus (TIDM) 39
type 2 diabetes mellitus (T2DM) 34–35, 60–62, *61*

United Nation's Children Fund (UNICEF) 58
urogenital microbiome, human 72–73
uterine transplantation (UT x) 89

vaginal defined medium with peptone (VDMP), in *Lactobacillus Plantarum* KCA1 129, *130*
vaginal infection 41–42
vaginal microbiome 32–33
  during pregnancy 102
vaginal microbiota transplantation (VMT) 87–89
Vogel, M.A. 139

Wade, P.A. 109
Wang, Z.K. 90
westernization, Nigerian diet 59–60

Zhang, Z. 110